I0484860

About the National Science and Technology Council

The National Science and Technology Council (NSTC) is the principal means by which the Executive Branch coordinates science and technology policy across the diverse entities that make up the Federal research and development enterprise. A primary objective of the NSTC is establishing clear national goals for Federal science and technology investments. The NSTC prepares research and development strategies that are coordinated across Federal agencies to form investment packages aimed at accomplishing multiple national goals. The NSTC's work is organized under five committees: (1) Environment, Natural Resources, and Sustainability; (2) Homeland and National Security; (3) Science, Technology, Engineering, and Math (STEM) Education; (4) Science; and (5) Technology. Each of these committees oversees subcommittees and working groups focused on different aspects of science and technology. More information is available at www.WhiteHouse.gov/ostp/nstc.

About the Office of Science and Technology Policy

The Office of Science and Technology Policy (OSTP) was established by the National Science and Technology Policy, Organization, and Priorities Act of 1976. OSTP's responsibilities include advising the President in policy formulation and budget development on questions in which science and technology are important elements; articulating the President's science and technology policy and programs; and fostering strong partnerships among Federal, state, and local governments and the scientific communities in industry and academia. The Director of OSTP also serves as Assistant to the President for Science and Technology and manages the NSTC. More information is available at www.WhiteHouse.gov/ostp.

About the Subcommittee on the Materials Genome Initiative

The Subcommittee on the Materials Genome Initiative (SMGI) contributes to the activities of NSTC's Committee on Technology (CoT). SMGI's purpose is to advise and assist the NSTC and OSTP on policies, procedures, and plans related to the goals of the Materials Genome Initiative (MGI). As such, and to the extent permitted by law, the SMGI defines and coordinates Federal efforts in support of the goals of MGI and identifies policies that will accelerate deployment of advanced materials. SMGI also tracks national priority needs that would benefit from MGI, identifies extramural activities that connect to MGI goals, and explores ways the Federal Government can advance the development of the Materials Innovation Infrastructure.

Copyright Information

Printed in the United States of America, 2014.

MATERIALS GENOME INITIATIVE

STRATEGIC PLAN

Materials Genome Initiative
National Science and Technology Council
Committee on Technology
Subcommittee on the Materials Genome Initiative

DECEMBER 2014

December 4, 2014

Dear Colleagues:

I am pleased to transmit to you the 2014 Materials Genome Initiative (MGI) Strategic Plan. MGI is an effort to double the pace of advanced-materials discovery, innovation, manufacture, and commercialization. It is part of a broader set of concrete actions launched by President Obama in 2011 to revitalize American manufacturing and the Initiative has made tremendous progress in its first three years. This Strategic Plan will serve as a framework for Federal agencies as they work to execute on the goals of MGI. The Strategic Plan highlights four sets of goals of the MGI:

(1) Leading a culture shift in materials-science research to encourage and facilitate an integrated team approach;
(2) Integrating experiment, computation, and theory and equipping the materials community with advanced tools and techniques;
(3) Making digital data accessible; and
(4) Creating a world-class materials-science and engineering workforce that is trained for careers in academia or industry.

This Strategic Plan was prepared by the Subcommittee on the Materials Genome Initiative (SMGI) of the National Science and Technology Council's Committee on Technology. The Subcommittee received and incorporated substantial input from a diverse array of stakeholders across the materials-science and engineering research and development community, including feedback gathered from thought leaders convened at multiple workshops in 2012-2013 and public comments received on the draft Strategic Plan earlier this year.

Achieving the MGI's vision of more rapid discovery and deployment of advanced materials will help ensure America's sustained leadership across the many sectors that utilize these technologies, including energy, electronics, national defense, and health care. I thank the SMGI for its hard work in developing this Strategic Plan and look forward to continued progress toward keeping our Nation on the cutting-edge of innovation in the advanced-materials domain.

Sincerely,

John P. Holdren
Assistant to the President for Science and Technology
Director, Office of Science and Technology Policy

NATIONAL SCIENCE AND TECHNOLOGY COUNCIL
COMMITTEE ON TECHNOLOGY
SUBCOMMITTEE ON THE MATERIALS GENOME INITIATIVE

National Science and Technology Council

Chair

John P. Holdren
Assistant to the President for Science and Technology
Director, Office of Science and Technology Policy

Staff

Jayne Morrow
Executive Director

Committee on Technology

Chair

Thomas Kalil
Deputy Director for Technology and Innovation
Office of Science and Technology Policy

Staff

Randy Paris
Executive Secretary

Subcommittee on the Materials Genome Initiative

Co-Chairs

Cyrus Wadia
Assistant Director, Clean Energy and Materials R&D
Office of Science and Technology Policy

Laurie Locascio
Director, Material Measurement Laboratory
National Institute of Standards and Technology
Department of Commerce

Harriet Kung (through Sept. 24, 2013)
Associate Director of Science for
    Basic Energy Sciences
Department of Energy

Linda Horton (beginning Sept. 24, 2013)
Director, Materials Sciences and
    Engineering Division
Office of Basic Energy Sciences
Department of Energy

Executive Secretary

James Warren
Technical Program Director for Materials Genomics
Material Measurement Laboratory
National Institute of Standards and Technology
Department of Commerce

Strategic Plan Development and Writing Team

Meredith Drosback	Office of Science and Technology Policy, Writing Lead
Julie Christodoulou	Office of Naval Research, Department of Defense
Mary Galvin	National Science Foundation
David Hardy	Office of Energy Efficiency and Renewable Energy, Department of Energy
Linda Horton	Office of Basic Energy Sciences, Department of Energy
William Joost	Office of Energy Efficiency and Renewable Energy, Department of Energy
Harriet Kung	Office of Basic Energy Sciences, Department of Energy
Laurie Locascio	National Institute of Standards and Technology, Department of Commerce
Bryan Morreale	Office of Fossil Energy's National Energy Technology Laboratory, Department of Energy
Harry Partridge	National Aeronautics and Space Administration
Charles Ward	Air Force Research Laboratory, Department of Defense
James Warren	National Institute of Standards and Technology, Department of Commerce

Executive Summary

Vision: *Advanced materials are essential to economic security and human well-being and have applications in multiple industries, including those aimed at addressing challenges in clean energy, national security, and human welfare. To meet these challenges, the Materials Genome Initiative will enable discovery, development, manufacturing, and deployment of advanced materials at least twice as fast as possible today, at a fraction of the cost.*

In June 2011, President Barack Obama launched the Materials Genome Initiative (MGI) alongside the Advanced Manufacturing Partnership to help businesses discover, develop, and deploy new materials twice as fast. For many years, the United States has been a dominant player in the discovery of transformative materials that are the basis of entirely new products and industries, yet the time lag between discovery of advanced materials and their use in commercial products can be 20 years or more. MGI will help position the United States for sustained leadership across the many sectors that utilize advanced materials from energy to electronics and defense to health care. MGI aims to capitalize on recent breakthroughs in materials modeling, theory, and data mining to significantly accelerate discovery and deployment of advanced materials while decreasing their cost. At the heart of MGI is the Materials Innovation Infrastructure, a framework of seamlessly integrated advanced modeling, data, and experimental tools that will be used to attain the MGI vision. Going beyond tools and techniques, MGI aims to link together networks of scientists spanning academia, National and Federal laboratories, and industry to more effectively share the information that underpins new material discovery and product development, and enables technological leaps.

Achieving this vision requires successfully addressing four key challenges:

(1) <u>Leading a culture shift in materials research</u> to encourage and facilitate an integrated team approach that links computation, data, and experiment and crosses boundaries between academia, National and Federal laboratories, and industry;

(2) <u>Integrating experiment, computation, and theory</u> and equipping the materials community with the advanced tools and techniques to work across materials classes and along the materials development continuum from research to industrial application;

(3) <u>Making digital data accessible</u> including combining data from experiment and computation into a searchable materials data infrastructure and encouraging researchers to make their data available to others;

(4) <u>Creating a world-class materials workforce</u> that is trained for careers in academia or industry, including high-tech manufacturing jobs.

The Federal agencies participating in MGI developed this document to outline the near-term steps the Federal government will take to achieve the vision put forth by MGI. The plan also describes the scientific and technical challenges identified by experts from the academic and industrial materials science and engineering communities that impede progress in nine materials classes and application domains that MGI can help address. The tools and scientific cultural evolution emerging from MGI can be directly applied to overcoming these grand challenges, and others that will emerge in the future, to meet the President's directive for more rapid discovery and deployment of advanced materials. Achieving these goals will be crucial to competitiveness in the 21st century and will help ensure that the United States maintains global leadership in innovation of emerging materials technologies in a wide range of industrial sectors including health, defense, and energy.

Table of Contents

Figures and Table

Introduction

Materials matter. The efficiencies of high-temperature turbine engines, biocompatibility of replacement joints and implants, operational life of advanced batteries, and sophisticated electronics that enable our digital world are all determined by the materials selected and optimized for the application. These innovations and myriad others shape the world we know and enable the future we envision. Yet transitioning a new material from initial discovery to practical use frequently takes 20 years or more.

The Nation's economic competitiveness and prosperity in the coming decades will depend critically upon the pace of American innovation. Recognizing the importance of advanced materials in supporting an innovation-driven U.S. manufacturing sector, President Barack Obama introduced the Materials Genome Initiative (MGI) in June 2011 with this aim: discover, develop, manufacture, and deploy advanced materials twice as fast, at a fraction of the cost.

This ambitious goal is within reach. Research conducted in the early 2000s demonstrated that a systems-level approach to material design, optimization, and implementation could significantly reduce design time and cost while improving quality. Some of these successes were chronicled in the 2008 National Research Council study *Integrated Computational Materials Engineering (ICME): A Transformative Discipline for Improved Competitiveness and National Security*.[1] One early example is the collaborative work of two aerospace engine design companies under the Defense Advanced Research Projects Agency (DARPA) Accelerated Insertion of Materials program. New principles of concurrently optimizing both design and manufacturing process enabled a new rotor disk design that had a 21% reduction in weight and 19% increase in burst strength, all achieved in nearly half the time of a typical development cycle.

Another early success story from 2007 was a new diesel engine rapidly brought to market through the use of modeling and analysis tools. This simulation-driven approach reduced development time and cost by decreasing the reliance on expensive and time-intensive hardware testing and by minimizing costly redesigns; it improved engine performance by allowing engineers to consider a broad range of design alternatives computationally, without the investment in hardware.[2,3] The work took advantage of foundational combustion modeling and laser diagnostics from the Department of Energy's Combustion Research Facility and this coupling of powerful computation with advanced characterization tools reflects an early example of the promise of MGI.

---

[1] National Research Council, *Integrated Computational Materials Engineering: A Transformational Discipline for Improved Competitiveness and National Security*, (2008) Washington, D.C.: The National Academies Press. (available online at http://www.nap.edu/catalog.php?record_id=12199).

[2] *Computational Materials Science and Chemistry: Accelerating Discovery and Innovation through Simulation-Based Engineering and Science*, Report of the Department of Energy workshop; July 26-27, 2010. (available online at http://science.energy.gov/~/media/ascr/pdf/program-documents/docs/Cmsc_rpt.pdf).

[3] Tickel, B., *Getting it Right the First Time*, ANSYS Advantage, 1, 3, 10 (2007). (available online at http://www.ansys.com/staticassets/ANSYS/staticassets/resourcelibrary/article/AA-V1-I3-Getting-It-Right-the-First-Time.pdf).

Innovative experimental tools also have a critical role in accelerating materials discovery and deployment. High-throughput experimental techniques have been deployed successfully in the field of pharmaceutical research to compress time to market for new drug therapies. In one comparison, combinatorial approaches were found to generate 1000 times more compounds with potential medicinal value than traditional methods for at least 600 times less cost per compound.[4] Over time, combinatorial techniques have expanded to other fields (e.g., catalysis, thermoelectric materials, and alloy design) and include both combinatorial synthesis and characterization, enabling rapid assessment and analysis.

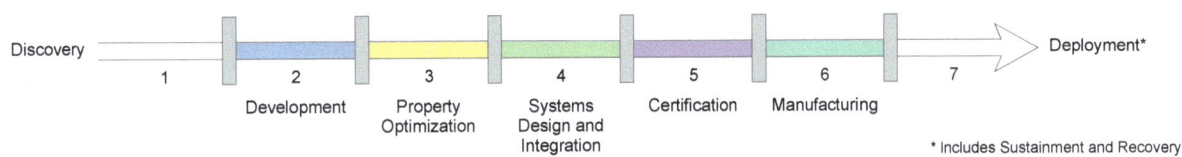

Figure 1. Materials Development Continuum

The successes and lessons learned from this early work illustrate the capabilities of different approaches as well as the potential for dramatic changes in workflow across all stages in the materials development continuum (see Figure 1) to accelerate materials to market and contribute to the design and goals of MGI.[5,6] The 2011 MGI white paper, *Materials Genome Initiative for Global Competitiveness*, described a Materials Innovation Infrastructure encompassing advanced computational, experimental, and data informatics tools (see Figure 2), along with a collaborative, integrated research paradigm for materials science and engineering.[7,8] The "MGI approach" seeks to uniquely and seamlessly integrate computation, experiment, and data to fuel the successful discovery of new materials and their more rapid deployment and incorporation into manufactured products.

Although MGI itself is a bold initiative, it is also inherently linked to other Administration priorities and Federal activities focused on addressing some of the Nation's most pressing needs in areas such as clean energy, national security, and human health and welfare, all of which have underlying challenges whose solutions require advanced materials. The President demonstrated the connection between MGI and other major Federal efforts intended to renew and revitalize U.S. manufacturing by launching MGI alongside the Advanced Manufacturing Partnership (AMP), a collaboration across government, industry, and academia to identify the most pressing challenges and transformative opportunities for improving technologies, processes, and products across multiple manufacturing industries. Additionally, MGI has a

---

[4] Persidis, A. "Combinatorial chemistry" *Nature Biotechnology* 18 (2000) IT50-52 (available online at http://www.nature.com/nbt/journal/v18/n10s/full/nbt1000_IT50.html).

[5] *Ibid*. 1.

[6] National Science Foundation, *Inventing a New America through Discovery and Innovation in Science, Engineering, and Medicine: A Vision for Research and Development in Simulation-Based Engineering and Science in the Next Decade*, (2010) (available online at http://www.nsf.gov/mps/ResearchDirectionsWorkshop2010/RWD-color-FINAL-usletter_2010-07-16.pdf).

[7] National Science and Technology Council, *Materials Genome Initiative for Global Competitiveness*, (2011) (available online at http://www.whitehouse.gov/sites/default/files/microsites/ostp/materials_genome_initiative-final.pdf).

[8] Throughout this document, the phrases "materials science and engineering" and "materials community" are meant to encompass a diverse range of disciplines relevant to the discovery, development, and deployment of advanced materials, including physics, chemistry, engineering design, an array of engineering disciplines, and manufacturing.

clear directive to provide an infrastructure for data sharing and access, a task in direct support of the 2013 Office of Science and Technology Policy memorandum on open data access for federally funded scientific research.[9] Further, MGI is closely linked to the National Nanotechnology Initiative (NNI) as materials scientists and engineers harness the advances in the understanding and control of material at the nanoscale made over the last decade due to NNI. When combined with these other initiatives and priorities, MGI has the potential to support the next wave of U.S. manufacturing and foster the kinds of cross-sector and cross-disciplinary collaborations that will open brand new avenues for innovation in efficiently solving national challenges.

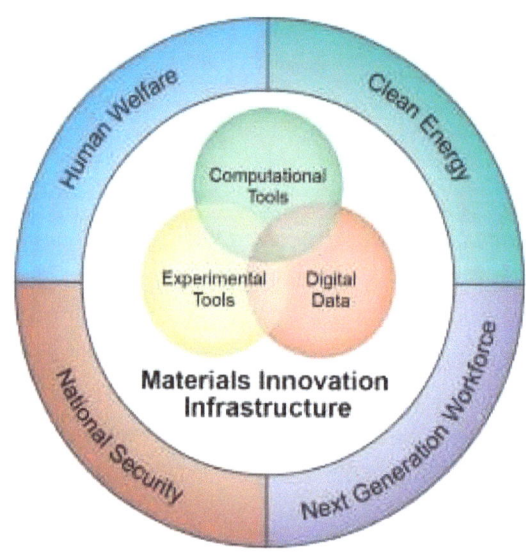

Figure 2. The Materials Innovation Infrastructure

MGI issues a unique challenge to the entire materials community: Deliver the next generation of materials into products in half the time at a fraction of the cost. This approach could lead to the accelerated discovery of new materials not yet possible with conventional tools, more rapid insertion of known materials into new products, and development of new technologies only imagined today (e.g., strong and dynamic-impact damping materials for military vehicles, helmets, and personnel armor; an ultra-lightweight material for cars that easily withstands high-impact crashes; or a thin-film battery material for cell phones that remain charged for weeks). The strategy described in this document (developed by Federal agencies with input from critical stakeholders from academia, National and Federal laboratories, and industry) is intended to guide and coordinate Federal activities and provide a clear technical path for carrying out the President's vision.

---

[9] Memorandum for the Heads of Executive Departments and Agencies from John P. Holdren, Director of the Office of Science and Technology Policy, on Increasing Access to the Results of Federally Funded Scientific Research (available online at www.WhiteHouse.gov/sites/default/files/microsites/ostp/ostp_public_access_memo_2013.pdf).

The next two chapters outline four strategic challenges to achieving the vision outlined by MGI, followed by a series of goals and objectives for successfully addressing these challenges. A subsequent chapter on achieving national objectives discusses how MGI can be leveraged to ensure that national needs are met in security, human health and welfare, clean energy, and infrastructure and consumer goods. Finally, a series of science and technology challenges from across the materials class and application spectrum are discussed. The tools and scientific cultural evolution that will develop as part of MGI can be directly applied to overcoming these challenges and others yet unidentified to meet the President's directive for more rapid discovery and deployment of advanced materials.

Key Challenges

Four key challenges have been identified as barriers between the current materials science and engineering paradigm and the future as envisioned by the Materials Genome Initiative (MGI). Summarized below, these challenges are: (1) a culture shift in materials research, development, and deployment; (2) integration of experiments, computation, and theory; (3) access to digital data; and (4) a well-equipped workforce. The goals and objectives outlined in the next chapter are designed to address each of these challenges through concerted efforts of public- and private-sector MGI stakeholders.

A Culture Shift in Materials Research, Development, and Deployment

Deeper integration of experiment, computation, and theory, as well as the routine use of accessible digital materials data, represents a shift in the usual way research is conducted in materials science and engineering. A major challenge facing MGI is how to establish mechanisms that will facilitate a flow of knowledge across the materials development continuum through deeper collaborations not only between theorists and experimentalists, but among academia, National and Federal laboratories, and industry as well.

Integration of Experiments, Computation, and Theory

A key characteristic that defines efforts in support of MGI is an integrated, collaborative workflow that draws simultaneously from experiments, computation, and theory. The vast spans of length and timescales covered by materials research create unique challenges for delivering quantitative and predictive scientific and engineering tools.[10] Important components of the Materials Innovation Infrastructure will be the development of advanced simulation tools that are validated through experimental data, networks to share useful modeling and analysis code, and access to quantitative synthesis[11] and characterization tools.

Access to Digital Data

Creating a digital data infrastructure that not only stores a wide range of data but is easily and reliably searchable is a challenge faced by many scientific disciplines, including materials science and engineering. Challenges facing the materials community across all technology readiness levels[12] include making users aware of the tools and data available; defining and implementing a widely accepted governance structure;

---

[10] Length scales can span from the size of atoms to physical structures common to everyday life, such as circuit boards, automobiles, and buildings; temporal scales can range from the fractions of a second characteristic of atomic interactions to the decades- or centuries-long lifetime of a manufactured object.

[11] Throughout this document, "synthesis" of a material is meant to include techniques using chemical reactions as well as techniques often referred to as fabrication, processing, or manufacture of materials.

[12] For examples and definitions of technology readiness levels see documentation from NASA:
http://www.hq.nasa.gov/office/codeq/trl/trl.pdf or DOD:
http://www.acq.osd.mil/chieftechnologist/publications/docs/TRA2011.pdf.

balancing security requirements with data usability and discoverability; and generating standards for describing data and assessing data quality. Meeting the vision of MGI will require broad and open access to validated data and tools generated by the materials community across the materials development continuum to allow both the reuse of individual data sets and the application of data analytics techniques to examine the aggregation of large volumes of data from many disparate sources.

A Well-Equipped Workforce

Even with development of a broadly accessible data infrastructure and new tools integrating experiment, computation, theory, and data, the next generation of scientists and engineers engaged in materials discovery, development, and deployment must be able to expertly use these tools to achieve the success promised by MGI. This challenge will be met in part through formal education in the application of this integrated approach for undergraduate and graduate students who will pursue careers in industry, National and Federal laboratories, and academia. For professionals already in the workplace, additional training may enable the widespread use of new tools and research methods. Also, before the future generation workforce can be equipped to take advantage of the Materials Innovation Infrastructure, instructors must first be provided information on these new tools, research approaches, and their value.

Strategic Goals and Objectives

The success of the Materials Genome Initiative (MGI) will be achieved by meeting the following four goals:

1. Enable a Paradigm Shift in Culture
2. Integrate Experiments, Computation, and Theory
3. Facilitate Access to Materials Data
4. Equip the Next-Generation Materials Workforce

This chapter expands on the substance of each of these goals and details specific objectives and milestones that will move MGI toward its aim of accelerating the development of new materials to meet national needs. Throughout this section, each milestone will include a list of agencies or interagency groups taking a lead role in executing the task.

In developing and executing the MGI activities described here, evaluation techniques and approaches will also be developed that allow assessment of both program efficacy and impacts. The details of the evaluation components of MGI remain undefined to date, but should include gathering sufficient project data to document what works well, the scientific output, and measures of increased pace and commercialization of materials innovation attributable to MGI.

Goal 1: Enable a Paradigm Shift in Culture

To achieve the vision of decreasing the time and cost of the materials discovery to deployment process, MGI must drive a shift in the way the community conducts research and development (R&D) and the commercial activities that produce and use materials. Fundamentally, this paradigm shift requires a change in the way teams collaborate. Collaboration today is widespread and productive, yet often narrowly confined to teams of scientists with similar expertise in theory, experiment, or simulation. Collaboration can become more fruitful through the seamless integration of theory; materials characterization, synthesis, and processing; and computational modeling. Further, advances in fundamental scientific knowledge and tools must be transitioned and integrated into engineering practice and application. This multidisciplinary approach will accelerate progress as results from each aspect inform the work of the others, enhancing communication across disciplines, avoiding delays and missteps, and enabling optimization.

This change requires engaging the entire materials community, from discovery through deployment, across the many engineering and scientific disciplines, academic departments, and industries that participate in activities related to materials. In addition, such a paradigm shift encompasses the development of this new collaboration model integrating theory, modeling, and experiment throughout the entire R&D continuum, from fundamental research through the design, optimization, and manufacturing phases. Therefore, industry plays a particularly important role in the strategy to form and adopt this new paradigm.

Objective 1.1: Encourage and Facilitate Integrated R&D

Integration across many domains is a cornerstone of the culture and techniques developed under MGI. Connections among theory, computation, and experiment; academia, National and Federal laboratories, and industry; science and engineering disciplines; and even Federal agencies are all critical to achieving the vision and demonstrating the value of the MGI approach. Successfully integrated research programs need strong multidisciplinary teams that span materials research activities. Communication within and among teams and across material and application domains is also a key component.

Attempts to demonstrate the value of this new collaborative, iterative structure have already begun. For example, the National Science Foundation's MGI program, *Designing Materials to Revolutionize and Engineer our Future (DMREF)*, emphasizes integration of computation and experiment in an iterative manner and encourages proposal evaluation on the basis of this collaborative research mechanism. This program and similar efforts ongoing at other Federal agencies have produced a small, but growing cohort of researchers that are using the iterative, collaborative MGI paradigm within their own research groups and with extended research partners. The Federal Government can support further transition to a research culture that includes integration across disciplines, as well as between the academic and industrial R&D communities, by emphasizing targeted support for this kind of work and bringing MGI elements into existing materials science and engineering R&D programs as appropriate. In fiscal year 2014 NSF added a third class of DMREF awardees to the existing group of scientists already supported by NSF, Department of Energy (DOE), and Department of Defense (DOD) MGI awards; each year more scientists become actively engaged in MGI-related projects and continuing to increase this number will facilitate more widespread development and adoption of the collaborative, integrated work style envisioned by MGI.

> **Milestone 1.1.1**: Over a two-year period, increase the cumulative number of researchers who have participated in MGI-related projects by 50%. [Department of Defense (DOD), Department of Energy (DOE), and National Science Foundation (NSF)][13]
>
> **Milestone 1.1.2:** Hold regular, multiagency principal investigator meetings to build a stronger MGI community. Include industry representatives in these meetings. [DOD, DOE, and NSF]

The Federal Government can further emphasize integration among academia, National and Federal laboratories, and industry by supporting activities that increase interactions between the communities. Examples include establishing new partnership opportunities around foundational engineering problems (FEPs), wherein an integrated, multidisciplinary team applies computational and experimental techniques toward achieving a specific performance goal in an engineering material or component.[14] Initially recommended by the 2008 National Research Council study, a FEP aids in research prioritization and demonstrates the power of integrated computational and experimental techniques. Partnerships between academic research and industry are critical for a shared understanding of which computational and experimental tools are needed most urgently, introduction and permeation of such tools, and training and education of the next-generation workforce required to use them.

---

[13] Throughout the remainder of this document, each milestone will list in brackets the agencies or interagency groups taking a lead role in the task.

[14] *Ibid.* 1.

Milestone 1.1.3: Over a two-year period, add multiple FEP projects supported by the Federal Government. [DOD and DOE]

Air Force Research Laboratory Foundational Engineering Problem in Composites

Fully realizing the potential of advanced polymer matrix composites (PMCs) in aerospace systems is limited by the lack of integrated simulation tools that capture enough detail to adequately represent the complexity of these high-performance materials in system designs. Specifically, the ability to link the chemistry of PMC processing with mechanical performance, particularly the load response and damage evolution for high-temperature PMCs, is hindering applications. The current design process typically relies on repetitive analysis and testing to incrementally build confidence in composite performance. This process results in overly conservative or inadequate component designs for complex structures and requires more time and higher testing costs.

The Air Force Research Laboratory's Materials and Manufacturing Directorate is leading a collaboration among General Electric, Lockheed Martin, Autodesk, Convergent Materials, University of Dayton Research Institute, and University of Michigan to develop the integrated materials engineering computational tools needed to model the complexity of PMCs across different spatial and temporal domains. This new work integrates high-fidelity processing and mechanics simulation tools for high-temperature PMCs into the composite material design, qualification, and certification processes. The resulting tools can be used for designing prototypical components such as an airframe wing box and an engine bypass duct to demonstrate reduced cost, time, and risk in using PMC materials. Additionally, reduced conservatism in designs and accelerated transition to next-generation materials will enable performance improvements and significant fuel savings for new aircraft.

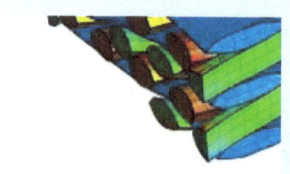

Image courtesy Air Force Research Laboratory

With broad Federal agency involvement in MGI, there are growing opportunities for cross-agency collaboration to take advantage of agency-specific expertise. For example, in 2013 DOE's Office of Energy Efficiency and Renewable Energy (EERE) awarded the first grants in a pioneering partnership between EERE and the National Institute of Standards and Technology (NIST). Under this program, NIST will curate repositories of materials data and models that result from research funded by the DOE-EERE program in lightweight automotive materials. This partnership can be modeled for extension to other agencies and can be applied to the broader MGI community through the dissemination of NIST-developed best practices in data management.

> **Milestone 1.1.4:** Over a two-year period, identify opportunities for three new MGI-related cross-agency grants or coordinated projects. [DOD, DOE, and NIST]

Objective 1.2: Facilitate Adoption of the MGI Approach

Supporting higher levels of collaboration solely through Federal investments will not be enough to realize the benefits of the MGI approach; long-term success will require building on these capabilities and new institutional incentives for broader adoption of MGI approaches for materials science and engineering research in academia, National and Federal laboratories, and industry. Ultimately, individual industrial sectors have to see the value in adopting this paradigm of collaboration. Targeted outreach to professional societies, industry consortia, and materials industry leaders can help to establish familiarity and stimulate discussion in the community. The National Science and Technology Council's Subcommittee on the Materials Genome Initiative (SMGI) will continue to serve as a convening agent to help facilitate interaction with industry and crystallize the vision of MGI.

Integration of new tools into the evolving industry and manufacturing landscape will be critical to the long-term success of MGI. Activity to establish the foundation and basic infrastructure building blocks of the MGI constitute the majority of this strategic plan but the SMGI acknowledges future work and leadership will be required to forge deeper integration and ties to industry. Such activity may include further development of public private partnerships to address advanced manufacturing needs as well as time-to-market constraints such as qualification and certification in the commercialization of new materials, devices, and systems.

Further, to facilitate exchange across academia, National and Federal laboratories, and industry and to facilitate the use of an MGI approach where applicable in industry, the Federal Government and the private sector could explore opportunities to support entrepreneurial training and industry experiences for students in all materials science and engineering-related disciplines. This type of educational program provides at least two benefits: the up-and-coming workforce has hands-on opportunities for applying MGI techniques learned in the classroom, and these students perform informal technology transfer by bringing expertise in the cutting-edge tools emerging from the research community directly to industry.

> **Milestone 1.2.1:** Work with materials science and engineering university programs, professional societies, and industry to define venues that promote interactions, transition, and integration between academic and industry researchers, including students, on MGI-related projects. [Subcommittee on the Materials Genome Initiative (SMGI)]

In addition, the Federal Government has demonstrated success in recent years in the use of incentive prizes and challenges to stimulate interest in well-defined R&D challenges; both the private sector and the Federal Government have available mechanisms through which to issue incentive prizes or challenges to solve identified technical challenges and to foster new collaborations.

Milestone 1.2.2: Over a two-year period, launch an incentive prize focused on demonstrating the use of MGI techniques to rapidly deliver new materials. [DOE, the National Aeronautics and Space Administration (NASA), and NIST]

Objective 1.3: Engage with the International Community

Accelerating the pace of discovery and deployment of advanced materials systems is in the economic interests of both the United States and its international partners in science, technology, and innovation. Many nations have identified advanced materials as a driver for industrial leadership and innovation; closer collaboration on these issues will provide mutual benefit, stimulating economies and bringing new opportunities for innovative technologies. While Federal agencies individually pursue international collaborations to further their mission goals, SMGI also has taken steps to engage with the international materials science and engineering community. Through the State Department and ministerial meetings led by the Office of Science and Technology Policy (OSTP), numerous opportunities exist for discussions of topics such as mutually compatible data access and sharing policies for materials data and identification of critical research needs in specific industrial sectors. Ultimately, these discussions will help both U.S. and partner research communities better target resources toward bottlenecks in the process and identify specific opportunities to reduce the time to market.

Milestone 1.3.1: Continue to pursue opportunities for collaborations with international partners, participate in international forums for discussions of materials science R&D, and build on strengths of existing international partnerships. [SMGI]

Goal 2: Integrate Experiments, Computation, and Theory

MGI emphasizes integration of tools, theories, models, and data from basic scientific research with the processing, manufacturing, and deployment of materials. The Materials Innovation Infrastructure will enable this integration by providing access to digital resources that contain the property data of known materials as well as the computational and experimental tools to predict these characteristics for new and emerging materials. Example applications include using integrated tool sets to identify replacements for critical materials, and then translating these new materials into the production pipeline. Ultimately, seamless integration of fundamental, validated understanding can be incorporated into the simulation and modeling tools used for materials discovery, product and manufacturing designs, component life predictions, and informed maintenance protocols.

The objectives that follow address the parts of this integration process that have been identified to date. The related, but distinct topic of open data access and associated issues relating to large data repositories is summarized in Goal 3, p. 20.

Objective 2.1: Create a MGI Network of Resources

Many of the initial Federal activities in support of MGI have been investments in a growing cadre of researchers whose work contributes to the development of the Materials Innovation Infrastructure. Connecting these researchers to each other, as well as connecting the broader materials community, from discovery and design through manufacturing, to the array of available capabilities, is the next critical task in developing a nationwide network of resources for materials science and engineering R&D.

Researchers need access to experimental capabilities for materials synthesis and characterization, whether for validating predictive capabilities of computational models or for empirical experimentation. High-tech experimental capabilities are available nationwide, and information about these resources will be a useful tool for researchers applying the MGI approach. Furthermore, MGI stakeholders also need access to modeling and software tools that are experimentally validated and widely functional across multiple platforms and user communities. These tools should include models that address the length and timescales required for practical applications, namely the size and projected lifetimes of engineered devices, while still preserving the scientific knowledge developed at the shortest lengths and times that determine the behavior and physical properties of the materials. Fundamental, science-driven, and well-characterized computational models need to be integrated with application-focused codes for integrated design, verification, performance prediction and sustainment, and other uses. Enhancing communication and sharing of common enabling tools through a community network of code and software developers will accelerate the availability of these tools to a wider range of users. A key aspect of this objective is establishing a resource with information on ongoing efforts across the materials research community engaged in the development of experimental and computational tools[15].

> **Milestone 2.1.1:** Work with the materials community across the full materials development continuum to establish an information inventory, including contact information or web links, for openly available codes, software, and experimental capabilities for synthesis and characterization, as a resource for the community. [SMGI]

Since the community that develops models and software is often distinct from the community that can make productive use of them, MGI needs to establish a path forward for transforming research-grade code into robust and sufficiently user-friendly software that meets the needs of user communities in academic, National and Federal laboratory, and industrial settings. In addition, pathways should be developed to nurture nascent efforts for the long-term development and maintenance of code and software packages; cross-disciplinary research programs that include computer science, information technology, and materials science are one method being explored. The private sector also pursues relevant software development on its own, providing another opportunity for productive public-private partnerships. Engagement with the private sector may also include facilitating the discovery of federally supported research outcomes and exploration of new business opportunities for commercially-supported software.

Through networking activities, researchers can foster the development and understanding of the best and proven approaches to successfully evolve the required software. Material-specific networks can identify priorities for interoperability standards, define necessary documentation, and identify common software modules that cross multiple applications.

> **Milestone 2.1.2:** Establish a network of research groups focused on developing predictive software for structural materials.[16] Document lessons learned and best practices for use in launching an additional network for other material and application areas. [DOD, DOE, NIST, and NSF]

---

[15] Throughout this document, the term "computational tools" is intended to include the theory, simulation, algorithms, and software enabling the design, discovery, and deployment of materials.
[16] Structural materials were chosen as an early proof of principle given the existing community-building work underway.

NanoHUB as a Model for a MGI Software Network

The development and distribution of software tools and associated educational resources are an important component of the Materials Innovation Infrastructure. One successful approach that the Materials Genome Initiative could emulate is nanoHUB.org, an online nanotechnology simulation community developed and operated by the National Science Foundation's Network for Computational Nanotechnology at Purdue University. NanoHUB empowers a worldwide community via cloud-based scientific computing and educational resources, providing a library of over 3,300 seminars, tutorials, and teaching materials to an active community of 257,000 users worldwide. NanoHUB's impact on research is demonstrated by more than 1,030 citations in the scientific literature and over 6,000 secondary citations. Furthermore, nanoHUB makes more than 270 constantly evolving simulation and modeling tools universally accessible and useful via fully interactive sessions in the cloud. Some 12,500 users run more than 430,000 simulations annually without any software installation, simply by using a web browser. Additionally, nanoHUB simulations are used at more than 180 institutions in formal classroom training that has reached 19,000 students to date. The image below graphically depicts the 250,000 users participating in nanoHUB as of February 2013. Red dots indicate users of education materials; yellow dots indicate simulation users.

Objective 2.2: Enable Creation of Accurate, Reliable Simulations

Success for MGI will require expansion of the current theory, modeling, and simulation tools available to the materials research and engineering community. Activities across the Federal Government will address predictive design of specific materials with the goal of developing robust computational tools with well-characterized predictive capability across the R&D continuum, including both discovery and processing steps, and making these tools available to the broader community. Development of these computational tools also relies on the broad availability of computational resources; the Federal Government has a long tradition in developing, and making available to the research community, high performance computational facilities. MGI will be able to take substantial advantage of these resources.

New computational methods implemented in software must be verified against known solutions and developed in concert with experiments to validate the output. As outlined in the next objective, specialized experimental tools often are required to provide the data necessary for validation. In addition, the integration of these advanced computational tools into experimental designs will drive faster and more robust experimental results from materials discovery through testing and integration of components.

Specific technical barriers in simulation also impede substantial advancement in the field of materials. For example, the materials science and engineering community has long recognized the challenges of multiscale theory and modeling. Since a material's performance is influenced by dynamics encountered at all length scales—from the atomic to macroscale—effective material design requires the integration of models, as well as the information derived from models, from many length scales. Equally important are the needs to quantitatively characterize and model a material's evolution to capture phenomena over the timescales relevant to application targets for industrial use. Directed efforts within MGI can address these specific technical needs; community input is needed to define the major scientific and technical challenges for theory, modeling, and simulation for all material types.

> **Milestone 2.2.1:** Convene the materials community working across all technology readiness levels to identify major scientific and engineering challenges for theory, modeling, and simulation for different materials classes and associated cross-cutting methods and algorithms. Hold a workshop annually and publish an associated report with an evolving focus on different material types.[17] Projected topics to be addressed in the first four years include structural materials, magnetic materials, energy storage materials, and electronic materials. [SMGI]

---

[17] All workshops outlined in this document are intended to include a broad array of relevant stakeholders as participants and to publish a summary report outlining the recommendations of the participants in addressing the workshop charge.

The Materials Project

Advanced materials will define the next generation of clean, safe, and affordable energy storage and distribution technologies, and first-principles modeling is providing a strong tool for accelerating the discovery of novel chemistries. While these techniques have broad applications, researchers at the Massachusetts Institute of Technology (MIT) and Lawrence Berkeley National Laboratory (LBNL) are using these theoretically sound calculations to rapidly determine key attributes of materials for energy storage because very little is known about these chemistries.

When designing novel compounds for energy storage, predicting a material's crystal structure is crucial. Typically, this exercise is treated purely as a computational energy minimization problem, a strategy fraught with enormous difficulty. Yet the use of data mining tools on the large amount of experimental data available for crystal structures may enable "learning" the rules of nature more efficiently in a mathematical way, a process which then rapidly drives the computations toward a new compound's most likely crystal structure. Such information would be invaluable for materials design and optimization, because it allows the linkage of compositional changes to those of crystal structure. In a successful example of this approach, the MIT and LBNL teams have identified many hundreds of new oxide compounds, several of which function as lithium (Li) battery electrode materials.

The interplay among experiments is also particularly important in understanding how materials will perform. For example, the fascinating recent discovery of $Li_{10}GeP_2S_{12}$ (LGPS), a novel solid-state electrolyte with extremely high lithium conductivity, led researchers to claim that LGPS was stable over a five volt (V) range. Using the large amounts of computed phase stability data now available through the Materials Project (www.materialsproject.org), such claims can be compared rapidly against computations. The results indicate that while the lithium conductivity could be confirmed with computations, first-principles phase diagrams clearly predict an electrochemical voltage window of no more than three V. More importantly, the computations allowed exploration of the impact of minor changes in the composition that could increase affordability or decrease ionic conductivity. These predictions have since been confirmed experimentally, demonstrating the power of computations for rapidly evaluating new ideas emerging from experiments and targeting optimization directions with the most potential.

Objective 2.3: Improve Experimental Tools—From Materials Discovery through Deployment

Materials are typically hierarchical in structure, from the atomic to the macroscale. Such hierarchies pose formidable challenges for both experiments and simulation. Tools to measure changes in structure, chemistry, and properties that have advanced the understanding of materials are found at x-ray and neutron facilities and in laboratories for electron, ion, and laser spectroscopy. Equally critical tools for the synthesis and fabrication of many materials are now available with atomic-level control of composition and structure and have extensive diagnostics capabilities for monitoring processing. Even so, the "best" of these tools typically are limited to specific materials systems or to small quantities of materials (e.g., thin films and nanoparticles). Many of the best characterization techniques still rely on significant sample preparations that are extraordinarily time-consuming and may modify or destroy the structures associated with the most interesting properties. Thus, to generate experimental data and validate predictions from theory, modeling, and simulation, continued advances in experimental tools are needed.

One example of improved experimental capability showing early promise is high-throughput experimentation, a method of parallelizing experiments to run many tests at once, often through the use of automation and small scale samples. Further, rapid growth in the application of high-throughput, combinatorial synthesis techniques in which large numbers of materials are rapidly synthesized in arrays of materials with different molecular or elemental compositions has the potential to identify unexplored pathways for materials synthesis and generate large quantities of experimental data. These synthesis techniques must be partnered with comparable combinatorial characterization capabilities that can rapidly measure the relevant properties of the individual materials in the array in order to maximize the potential of high-throughput experimental methodologies for creating and identifying new materials with desirable properties.

The complexity of materials for today's technologies imposes additional challenges for MGI. For example, in advanced electronics and photonics, the material of interest is itself an interface between other materials or a surface that requires exquisite control of composition and doping for optimum performance. The ability to make materials with this level of structural control will require the development of new synthetic techniques and processes. A 2009 report from the National Research Council, *Frontiers in Crystalline Matter: From Discovery to Technology*, points to a national need to enhance the U.S. capability for making crystalline materials including two-dimensional and thin-film crystals.[18]

---

[18] National Research Council, *Frontiers in Crystalline Matter: From Discovery to Technology*, (2009) Washington, D.C.: The National Academies Press. (available online at http://www.nap.edu/catalog.php?record_id=12640).

Soft Materials Data Generation and Exchange Through the nSoft Consortium

An unparalleled range of properties—from fluidity to steel-like strength—can be achieved with soft materials, such as polymers, proteins, and colloids, simply by changing their molecular architecture and processing parameters. These unique materials are often suggested as an optimal solution for emerging societal needs in advanced body armor, lightweight transportation, sustainable agriculture, advanced energy storage and delivery, and the next generation of advanced therapeutics. Yet the complex relationship of molecular architecture, processing parameters, and performance of soft materials defies current characterization methods and challenges any attempt to develop predictive performance models.

Lacking this predictive modeling capability, many researchers are forced to adopt more costly or insufficient solutions to understand these materials. Neutrons, with their powerful ability to highlight individual molecules and phases, can be used to characterize materials with high precision under processing conditions, thus providing a way of obtaining the critical physical parameters needed for integration into state-of-the-art predictive modeling tools. The Materials Genome Initiative creates an opportunity to leverage unique data derived from both experiment and computation to foster a new generation of high-performance, low-density, cost-effective materials. Additional benefits could be realized in the stability of high-concentration antibody formulations, shear thickening fluids for body armor, and membranes for clean water technology. The National Institute of Standards and Technology (NIST) Material Measurement Laboratory is committed to providing these relationships through the nSoft industrial consortium (www.nist.gov/nsoft), which operates a suite of world-leading neutron-based measurement tools at the NIST Center for Neutron Research (NCNR). nSoft members are leading manufacturers of soft materials, spanning industrial sectors from petrochemicals to aerospace to biopharmaceuticals. In addition to providing critical data required for predictive modeling, the nSoft membership represents a key space for gaining tangible connections between stakeholder needs and transfer of data as well as identifying emerging trends in manufacturing.

Since the goal of MGI is to accelerate the discovery, design, development, and deployment of new materials into manufactured products, expanded use of real-time methods is essential for dynamic analysis of materials *in situ*—that is, taking measurements in realistic environments (not only in a vacuum or at ambient conditions) during the synthesis, processing, and "use" of materials. These types of data are necessary for validating the accuracy of theories and models, completing data sets where theories and models are not yet comprehensive, and informing predictions of how a material's properties emerge and change with time.

> **Milestone 2.3.1:** Convene a multiagency workshop to assess the current state and future directions for characterization tools that allow *in situ* and *in operando* assessments of materials properties, synthesis, and processes. [DOD, DOE, NASA, NIST, and NSF]

An MGI approach contributes to accelerated materials development and deployment, in part, by integrating computational and experimental tools to better predict, and therefore control, how manufacturing process parameters will affect final material and product performance. This capability allows the co-development of component design and optimized properties tailored for the product. Consideration of the full range of material characteristics, properties, and manufacturing steps that are required to produce a material or incorporate a material into products is integral to achieving the goal of

MGI. With structural materials, for example, manufacturing processes may include machining, forming, casting, and welding, as well as quality control to ensure that the materials achieve the desired final properties. For other types of materials such as catalysts, the materials may be the final product and a host of complex synthesis and processing steps result in a material with the required functionality. In all cases, materials development and implementation must be performed responsibly across the entire life cycle of the material; the use of scarce materials should be minimized, potential toxicity should be assessed early in the materials development process, and options for reuse or safe disposal of materials should be considered.

Once materials are deployed, prediction of their performance lifetime in service is crucial. The integrated tools developed under MGI to understand a material's lifetime behavior, and the associated uncertainties in those predictions, also will enable users to create designs for maintainability. In addition, there is substantial benefit to integrating diagnostic systems that allow for real-time awareness of a material's evolution (changes in structure and chemistry) and functional performance. MGI activities will include developing the computational and experimental tools for advancing today's understanding of how time and environmental factors impact a material's structural evolution.

The development of improved sensor systems, associated software for in-line quality assessments during manufacture, and reliable predictions of time to failure would substantially benefit many application areas. These post-deployment materials evolution challenges are rarely incorporated into the materials design paradigm because the models describing these processes are immature and thus, of limited utility in design. Such depth of understanding could enable accelerated tests of materials, further reducing the time for critical steps in materials development and product design, integration, and certification.

> **Milestone 2.3.2:** Convene a series of multiagency workshops to identify major scientific and technical challenges limiting the application of the integrated, collaborative MGI approach toward advanced manufacturing of materials and products. Conduct workshops in the first four years focusing on specific material classes and applications including lightweight metals, catalysts, batteries and energy storage, and semiconductors and integrated circuits. [NIST, DOE, DOD, and NSF]

Understanding the time required at each step in the materials development continuum, from materials discovery to deployment in the marketplace, is critical to decreasing the total time to market for new materials. Existing evidence is largely anecdotal, and studies are needed to benchmark the current state of the art across many industries, materials classes, and applications to be able to measure and assess success.

> **Milestone 2.3.3:** Initiate benchmarking studies to quantify the current time to market for a subset of materials classes or applications. [NIST]

Objective 2.4: Develop Data Analytics to Enhance the Value of Experimental and Computational Data

A growing challenge across many scientific domains is the magnitude of data—both computational and experimental—that can be produced with some of the current generation of tools. The next goal in this strategic plan discusses the objectives and milestones associated with developing and maintaining the required databases to enable assessments of this data. The availability of high-quality experimental and computational data also presents an opportunity for data mining and analysis to expand and accelerate discovery of new materials and predictions of materials with new functionalities. Data mining and analysis will be enabled by the availability of materials data in common formats and with consistent metadata to establish the information's provenance. In addition, some experimental results can be accelerated by real-time analysis of experimental data with modeling and simulation tools that enable data interpretation, guiding the evolution of ongoing experiments.

> **Milestone 2.4.1:** Convene a path-finding workshop focused on the status of computational tools for data analytics for applications emerging from materials sciences and engineering. [NIST]

Goal 3: Facilitate Access to Materials Data

The availability of high-quality materials data is crucial to achieving the advances proposed by MGI. Materials data can be used for input in modeling activities, as the medium for knowledge discovery, or as evidence for validating predictive theories and techniques. If made widely available, disparate sources of materials data also could be inventoried to identify gaps in available data and to limit redundancy in research efforts. To benefit from broadly accessible materials data, a culture of data sharing must accompany the construction of a modern materials data infrastructure that includes the software, hardware, and data standards necessary to enable discovery, access, and use of materials science and engineering data.

Driven by a diverse set of communities with unique and heterogeneous requirements, this data infrastructure should allow online access to materials data to provide information quickly and easily. A set of highly distributed repositories should be available to house, search, and curate materials data generated by both experiments and calculations. Community-developed standards should provide the format, metadata, data types, criteria for data inclusion and retirement, and protocols necessary for interoperability and seamless data transfer. This effort should include methods for capturing data, incorporating these methods into existing workflows, and developing and sharing workflows. This strategy requires a structured approach starting with the commissioning of path-finding efforts to identify the required architecture, standards, and policies needed to build a materials data infrastructure. Important to note is that many of the needed information technology solutions are available or under development; the strategy defined here leverages these technical advances and concentrates on applying them in the context of materials research.

Objective 3.1: Identify Best Practices for Implementation of a Materials Data Infrastructure

A materials data infrastructure combining the software, hardware, and community-wide standards to allow discovery, access, and use of materials data is one of the critical components of the Materials Innovation Infrastructure envisioned by MGI. The variety and complexity of materials data have hampered the creation of a single, widely accepted vision of the structure, organization, and other specifics needed for a materials data infrastructure. Given these complexities and the endeavor's scale, critical objectives are to explore best practices used by existing data collections and to learn from ongoing efforts to establish materials data repositories and other data infrastructures. In establishing best practices, lessons from similar efforts in other fields will be exploited. For example, the Human Genome Project of more than a decade ago created a revolution in the field of genomics that continues to be fueled by a consolidated data effort.[19] Likewise, the earth sciences community continues to explore and define the necessary elements of their shared data model through the path-finding EarthCube collaboration.[20]

> **Milestone 3.1.1:** Convene a series of multiagency workshops that engage stakeholders, including researchers from academia, industry, publishing, and government to establish the needs of the disparate materials communities, identify the barriers to creating a materials

---

[19] For more information, see www.ornl.gov/hgmis.
[20] For more information, see www.nsf.gov/geo/earthcube.

data infrastructure, and define potential methods of overcoming these obstacles. [DOD and NIST]

Best practice assessments will be coordinated across the Federal agencies to ensure that the outcomes meet agency missions while maximizing efficiency and efficacy of the resulting infrastructure. This coordination also will allow potential integration of the resulting infrastructure into other cyberinfrastructure activities within the Federal agencies, maximizing the benefit to a broader scientific community.

While assessing the various technical requirements associated with creating and maintaining repositories, the stakeholder community should identify needs associated with using the data, such as new tools to access information quickly and accurately as well as advanced data analytics. MGI activities will be able to leverage ongoing efforts by the Networking and Information Technology Research and Development Program (NITRD)[21] and the broader community surrounding Big Data[22] to provide some solutions to these questions.

Current agency data management plans, such as those pioneered by NSF[23], require researchers to consider how they will manage the data generated during the course of research and make that data available to other researchers. With the recent OSTP guidance[24] to increase access to the results of federally funded scientific research, data management will evolve and over time, more data will become publicly available.

> **Milestone 3.1.2:** Foster ongoing discussion of best practices in data management plans used by participating agencies with the opportunity to leverage these for broader applications within the MGI community. [SMGI]

An important means for incentivizing data sharing is to ensure that those who generate the data receive proper credit. Thus, community norms need to be developed for proper citation of digital data, including the technical infrastructure to make data citation straightforward and function in a manner similar to the digital object identifier system currently used to cite published papers. Numerous national and international bodies, such as the International Council for Science and National Information Standards Organization, are actively studying this topic and developing practices and standards for data attribution and citation that MGI-developed repositories could choose to adopt.[25,26,27] Over the long term, adoption

---

[21] For more information, see www.nitrd.gov.

[22] For more information, see www.WhiteHouse.gov/blog/2012/03/29/big-data-big-deal.

[23] For more information, see www.nsf.gov/bfa/dias/policy/dmp.jsp.

[24] For more information, see www.whitehouse.gov/sites/default/files/microsites/ostp/ostp_public_access_memo_2013.pdf.

[25] For example: International Council for Science: Committee on Data for Science and Technology, Data Citation Standards and Practices Task Group (available online at www.codata.org/taskgroups/TGdatacitation/index.html).

[26] For example: DataCite (available online at www.datacite.org).

[27] For example: National Information Standards Organization Forum, "Tracking it Back to the Source: Managing and Citing Research Data" (2011) (available online at www.niso.org/news/events/2012/tracking_it_back_to_the_source).

of data attribution and citation standards within materials science communities will require a combination of community dialogue, education, and adaptation.[28,29,30]

Objective 3.2: Support Creation of Accessible Materials Data Repositories

Objective 3.1 aims to identify the elements of a materials data infrastructure and associated standards necessary to support repository interoperability and seamless data transfer. This infrastructure is anticipated eventually to comprise a federation of public and participating private repositories (or "federated databases"), which may be networked together while remaining geographically separate, providing online access to materials data for both research and industrial applications. These highly distributed repositories would be available to house the curated data and incorporate data generated by both experiments and simulations. Yet several challenges remain in defining and creating the infrastructure within which these repositories would operate.

A successful data infrastructure will provide useful materials information to academia, National and Federal laboratories, and industry quickly and easily. Such an infrastructure should provide data together with sufficient descriptive information to properly identify it, assess its utility, and support both simple and complex semantic-based queries across the range of federated data repositories.

> **Milestone 3.2.1:** Develop and implement at least three materials data repository pilot projects to assess a range of repository models and initiate the definition of a materials data infrastructure model. [DOD, DOE, and NIST]

These pilot projects will be used to explore, adapt, and test the technological modalities needed to develop a data infrastructure. Communities of interest would conduct these pilot projects to define the standards requirements, including formats and protocols for data sharing and interoperability, for enabling a federated system without explicit central control. The end product would be a model of a working system comprising high-value and practical community-based standards, and it would demonstrate tools for search and identification of existing experimental or calculated materials data that could be used in a specific endeavor. Data would be presented with sufficient information to assess and select which data are useful, and appropriate linkages would be provided to the data access mechanisms.

---

[28] Nelson, B., 2009. "Data Sharing: Empty Archives," *Nature* 461, 160.
[29] Nature Editorial, 2013. "Disciplinary Action: How Scientists Share and Reuse Information Is Driven by Technology but Shaped by Discipline," *Nature*, 495, 409–410.
[30] For example: Research Data Alliance (available online at www.rd-alliance.org).

Goal 4: Equip the Next-Generation Materials Workforce

For the Nation and materials research community to take full advantage of the MGI framework outlined in previous sections, the next-generation materials workforce must be trained in these new research methods. Students will need access to an education that enables them to work productively in teams whose expertise covers the broad materials spectrum from synthesis and characterization, to theory and modeling, to design and manufacture. In practical terms, students who will go on to become experts in materials synthesis, processing, or manufacture, for example, must have enough training to understand materials modeling and theory, while modelers and theorists must understand the vocabulary and challenges of those who make, characterize, and implement materials solutions. Accomplishing this goal will require continued updates in the materials science and engineering curricula as well as in departments that contribute to educating the next-generation materials discovery, development, and deployment workforce. Just as many materials science and engineering departments have added computational materials science to their curriculum in recent years, formal instruction on data analytics, uncertainty quantification, and the integration of simulation, experiment, and theory will provide students with the foundation to successfully implement an MGI approach in their academic or industry careers.

The Federal government's broader activities in science, technology, engineering, and mathematics (STEM) education are driven by the *Federal Science, Technology, Engineering, and Mathematics (STEM) Education 5-year Strategic Plan*, which identifies five priority areas for STEM education investment.[31] Two of these priority areas, enhancing the STEM experiences of undergraduates and designing graduate education for tomorrow's STEM workforce, are pivotal for achieving the goals of MGI and the Federal government's specific activities will be designed to coordinate with the implementation strategies under development in these areas.

Objective 4.1: Pursue New Curriculum Development and Implementation

As a prelude to preparing students for working in a collaborative and iterative manner utilizing the tools developed under MGI, the first step is to educate faculty about the goals of MGI including its approach and tools. The Federal Government is enabling this process through support for numerous workshops and academic research grants funded by MGI programs at NSF, DOD, and DOE. For MGI to be successful, researchers will need to work closely in teams of professionals from disparate backgrounds. Researchers who focus on making or processing materials also, for example, must have the analytical expertise to understand the capabilities that modeling materials and processes can enable. Likewise, theorists and modelers must be exposed to topics such as the processes and limitations of making, processing, and characterizing materials.

> **Milestone 4.1.1:** Create opportunities, such as summer schools or laboratory internships, aimed at training faculty, postdoctoral researchers, and graduate students in the MGI approach to materials science and engineering. Topics may include familiarizing experimental materials scientists with current state-of-the-art modeling and theory and familiarizing

---

[31] National Science and Technology Council Committee on Education, *Federal Science, Technology, Engineering, and Mathematics (STEM) Education 5-year Strategic Plan* (2013) (available online at http://www.whitehouse.gov/sites/default/files/microsites/ostp/stem_stratplan_2013.pdf).

computational materials scientists with synthesis and characterization techniques and limitations. [DOD, DOE, and NSF]

As the number of faculty engaged in integrating theory, modeling, and experimentation increases, curriculum supporting this approach, both in materials science and engineering and other departments, will be developed. Materials research is inherently interdisciplinary with participation from experts beyond the traditional academic materials science and engineering departments, including but not limited to physics, chemistry, chemical engineering, bioengineering, applied mathematics, computer science, mechanical engineering, and manufacturing engineering. Therefore, the leadership of academic departments, universities, and professional societies will be crucial.

> **Milestone 4.1.2:** Convene university departments engaged in materials research, including physics, chemistry, bioscience, and engineering, to identify: (1) the educational approaches and the institutional and professional incentives needed to sustain interdisciplinary research, and (2) opportunities to better integrate theory, modeling, experimental, and data analytics training for undergraduate and graduate students pursuing careers or research in materials. Identify and share best practices through annual meetings of academic leaders. [SMGI]

The Federal Government can engage universities to facilitate development and adoption of new content and methods in related curricula through a number of potential mechanisms, including those covered in Milestone 4.1.2. NSF, the lead agency in implementing Federal STEM undergraduate and graduate education activities, would coordinate Federal efforts to foster curriculum development and implementation related to MGI goals.

Many undergraduate and graduate students studying materials science will pursue careers in industry where they will be responsible for developing and deploying the advanced materials of the future. For this reason, it is important to engage industrial leaders in identifying the skills and expertise that will enable the next generation of materials researchers to incorporate effective MGI-driven tools and practices in establishing a vibrant 21st century materials and manufacturing base in the United States. Input is needed from industry, National and Federal laboratories, and academia to address the evolving capability requirements and curriculum changes.

> **Milestone 4.1.3:** Facilitate discussions among Federal agencies, academia, and industry to identify capabilities and skill requirements for recent graduates entering the industrial workforce and ways to prioritize their development at educational institutions. [SMGI]

Fostering Education in MGI Techniques

Enabling the capabilities developed under the Materials Genome Initiative to be used widely and effectively to accelerate materials development requires equipping the next-generation workforce with new tools and multidisciplinary work experiences. While not the norm, one bold approach to providing undergraduates with such an environment has been developed in a series of materials design education innovations at Northwestern University.

Recognizing that an engineering discipline is defined by what can be practiced with a bachelor's degree, the Northwestern-led Steel Research Group design consortium developed a computational design methodology that can be taught to undergraduates, starting with an undergraduate Materials Design course in 1989.[32] In a unique integration of research and education, teams of materials science undergraduates conduct annual iterations of theoretical design optimization employing the newest experimental measurements and model and simulation predictions. The course features a series of labs teaching a suite of computational design tools grounded in the materials fundamental databases and the graphical parametric design integration approach.[33]

The undergraduate teams are coached by doctoral students participating in funded design projects.[34] These Ph.D. students are, in turn, assisted by a broader group of graduate students contributing to projects under a special interdisciplinary doctoral cluster program in Predictive Science and Engineering Design (PSED). A central outreach activity to promulgate the new design practices to a broader audience is a Master of Science certificate program in Integrated Computational Materials Engineering through which first-year M.S. students also participate in the interdisciplinary PSED seminar, culminating in an integrative project in the Materials Design course.

Under the materials science and engineering undergraduate program featuring multiyear design education, undergraduates taking the Materials Design course in their third year can participate in the experimental validation of their design prototypes in their senior projects the following year. To enhance recruitment to the materials program, student teams from a special "Murphy Scholars" section of a freshman-level Engineering Design and Communication course also collaborate with the undergraduate design teams, adding exploration of device applications for the new materials.[35] Featuring a highly effective "techmanities" cross-cultural design program, the latter course is co-taught by humanities faculty in the Writing Program. The broader goal is to develop, assess, and enable similarly new integrated approaches to engineering education across the expanded collection of materials classes.

Objective 4.2: Provide Opportunities for Integrated Research Experiences

Opportunities for students to become engaged in research with faculty or in industrial internships often augments science and engineering coursework. These activities cement the knowledge gained through

---

[32] Olson, G. B., 1991. "Materials Design: An Undergraduate Course," in P. K. Liaw, J. R. Weertman, H. L. Markus, and J. S. Santner (Eds.), *Morris E. Fine Symposium*, TMS-AIME, Warrendale, PA. 41.

[33] Olson, G. B., 2001. "Brains of Steel: Mind Melding with Materials," *International Journal of Engineering Education*, 17, 468.

[34] McKenna, A. F., Colgate, J. E. and Olson, G. B., 2006. "Characterizing the Mentoring Process for Developing Effective Design Engineers," *Proceedings of the American Society for Engineering Education (ASEE) Annual Conference*.

[35] McKenna, A. F., Colgate, J. E., Carr, S. H., and Olson, G. B., 2006. "IDEA: Formalizing the Foundation for an Engineering Design Education," *International Journal of Engineering Education* 22, 671.

coursework and expose students to the excitement of materials discovery and deployment in products via real-world, hands-on experience. Likewise, postdoctoral researchers can benefit from opportunities to expand their network of collaborators in academia, the National and Federal laboratories, and industry during this early-career training period. To hone their knowledge and skills, undergraduate students, graduate students, and postdoctoral researchers will need to practice MGI-related techniques in academic, government, and/or industrial laboratories as a standard part of their training. Industry will play a critical role in this activity, and a community-led workshop should consider appropriate roles of industry, Federal support, and new opportunities for mentoring activities related to MGI topics (e.g., seminars, internships, job shadowing, or capstone project evaluation).

> **Milestone 4.2.1:** Facilitate a dialogue on best practices and opportunities both for existing programs and potential new partnerships among industry, universities, Federal agencies, and National and Federal laboratories to provide opportunities for real-world experience in applying the MGI approach. [SMGI]

> **Milestone 4.2.2:** Develop and propose options for expanding postdoctoral research opportunities to include targeted positions in research teams specifically implementing the MGI approach. [SMGI]

Achieving National Objectives

Advanced materials will facilitate development of the disruptive technologies that will continuously improve the quality of life for future generations. To keep U.S. industry globally competitive in critical sectors such as national security, human health and welfare, clean energy, infrastructure, and consumer products, product innovation and manufacturing should occur more quickly and efficiently than comparable efforts by competitors. The Materials Genome Initiative (MGI) will provide an innovative technological and cultural framework that leverages integrated multidisciplinary research and engineering to span public, private, and academic sectors and successfully accelerate the improvement of existing materials and processes and development of the visionary materials of the future.

This chapter highlights the relevance of a successful MGI to achieving national objectives in security, human health and welfare, clean energy systems, and infrastructure and consumer products.

National Security

The Department of Defense (DOD), Department of Energy's National Nuclear Security Administration (NNSA), and national defense laboratories are significantly invested in materials research explicitly for national security. While DOD uses advanced materials to help protect and arm American troops, and NNSA uses advanced materials to ensure the safety and effectiveness of the American nuclear weapons deterrent, materials also play a role in many other areas of national security. Materials advances are important for lighter-weight protection systems and vehicles, advanced energetic materials, composites used in turbine engines, lifetime prediction of defense systems, electronics, and energy storage and distribution, among other applications. Many important materials developments eventually are transitioned into commercial products that enhance the well-being of the country at large.

Probing Fuel Cells *In Situ* with Raman Spectroscopy

Advanced fuel cell technologies offer highly efficient, clean, and quiet power generation. The portable systems envisioned for military applications also must be rugged and robust. Designs must presume austere conditions where fuel sources may be limited and not easily certified. Understanding the complex reaction kinetics associated with oxygen reduction and fuel oxidation occurring in solid-oxide fuel cell (SOFC) designs under specific operating conditions using a variety of fuels is key to providing dependable power sources.

Recognizing a critical need for quantitative data describing reactions under relevant operating conditions, the Office of Naval Research supported the development of advanced *in situ* characterization tools. A team from the U.S. Naval Research Laboratory and Montana State University has developed *in situ* optical and thermal diagnostics for probing SOFCs at typical operating temperatures of 700° to 800°C using Raman spectroscopy and thermal imaging techniques in combination to determine *in situ* chemistry and electrochemical reactions at the SOFC anode.[36] These noninvasive, nondestructive, real-time monitoring techniques provide quantitative data and visualization of complex phenomena. Tightly integrating the development of theoretical and predictive models with such advanced analysis both validates and informs more accurate models, enabling researchers to begin to predict how SOFC materials interact with different hydrocarbon and alcohol fuels while in operation.

For example, this diagnostic technique already is revealing conditions that exacerbate carbon production during cell operation or that limit detrimental effects on cell performance. As a result, SOFC performance could be improved through choice of fuel or SOFC materials composition and structural changes. Developing diagnostic capabilities for assessing proper performance functioning of components during operation is also possible.

Human Health and Welfare

Advanced materials are critical to the continuous provision of affordable, abundant, and environmentally responsible life essentials, including food, water, shelter, and healthcare commodities. For example, emerging biocompatible materials are likely to continue to play a crucial role in technology advancements for making targeted medical devices, smart prostheses, and cultivating artificial organs. Organic and solid-state sensors support medical diagnostic tools and *in vivo* pharmaceutical products delivery, and novel chemistries advance delivery and function of medications. New separation technologies enable broader access to clean drinking water, a major global health issue. Applying MGI principles to the development of these technologies will allow continued U.S. global leadership in providing quality of life for humanity.

---

[36] Pomfret, M. B., Walker, R. A., and Owrutsky, J. C., 2012. "High-Temperature Chemistry in Solid Oxide Fuel Cells: *In Situ* Optical Studies," *Journal of Physical Chemistry Letters* 3, 3053.

Clean Energy Systems

Although energy demand in the national energy portfolio is projected to observe only modest increases over the next 20 years, the equipment and tools used to support the energy infrastructure will change significantly. Acknowledging that global demand is expected to increase by about 50 percent in that same timeframe, the need to support rapid materials development is paramount if supply chains are to be maintained, especially for new technologies.

Within an "all-of-the-above" national energy strategy—including fossil, nuclear, and renewable sources to meet future energy demands—the discovery and deployment of advanced materials for harnessing, converting, distributing, and utilizing these energy sources are crucial for providing humanity with affordable, abundant, and environmentally responsible energy systems. Examples of such sustainable systems include innovative materials to more fully utilize the vast solar resources, pioneering energy-storage materials enabling a diverse energy harnessing and delivery infrastructure, novel alloys enabling efficient energy conversions in extreme environments, and groundbreaking catalysts promoting the production of energy-dense liquid fuels from a variety of feedstocks.

Designing Catalysts from First Principles

Catalysts are essential in the manufacturing of over 95% of industrial chemicals and fuels, because they make difficult conversions technically and economically feasible. Well-known commercial processes include, for example, ammonia synthesis with the Haber-Bosch process. Traditionally, catalysts for a specific conversion have been identified by a search guided by previous experience. When an untried conversion is needed, such searches can be lengthy and frequently are unsuccessful. Linking materials structure to reactivity for a certain type of chemical bond usually provides insufficient guidance, because the parameter space includes the specific reaction environment plus local, secondary, and long-range structures and their dynamics for both the catalysts and the reacting substances. This wide parameter space also includes interactions among the reactants, solvent, interfaces, subsurfaces, and bulk of materials, as well as excitation from various energy sources.

A completely *ab initio* design of catalysts for a given conversion without previous experience has yet to be achieved, but such a design is much closer to being feasible by means of rational approaches such as those envisioned by the Materials Genome Initiative. An example is the SUNCAT Center for Interface Science and Catalysis at the SLAC National Accelerator Laboratory where numerous activities are underway to improve understanding of catalyst behavior and properties. Electronic structure theory is used in combination with experimental methods to model surface reactivity. Use of advanced x-ray synchrotron sources at SLAC with synthesis facilities at Stanford University enables atomic-level resolution in structural data and molecular-level detail in mechanistic understanding. X-ray studies provide bonding information under the same conditions as the catalyst would experience in applications. Interfacial spectroscopy, in combination with theories of surface interactions, correlations of bonding trends, and simulations of surface dynamics, provides accurate quantification of energy distribution in space and time. Studies of yields and reactivity of materials exposed to full catalytic cycles provide correlations among structure, stability, and performance.

In parallel with these activities, methods for more predictive theories are being developed. These methods involve reexamination of electronic structure theories to maximize accuracy and minimize uncertainty. They also include intensive data management consistent with a hybrid set of data sources. This extensive combination of experimental and theoretical tools and approaches is crucial for enabling sought-after transformations, such as benign biomass depolymerization using light and inexpensive photocatalysts.

Infrastructure and Consumer Goods

In addition to the three sectors discussed previously, there are myriad other technology and infrastructure applications that contribute to the Nation's economic prosperity and continue to necessitate development of new materials. For example, longer-lasting, safer bridges and roadways may be enabled by advances in concrete designs. The next generation of cell phones could be built using flexible, solar-powered materials. Advanced optical fibers could one day provide even faster internet access. These applications and many more disruptive technologies not yet envisioned may be possible with the discoveries and new applications accelerated by MGI. Ultimately, deployment of materials into mass-produced consumer goods should be done with attention paid to the safety, disposal, and possible reuse of the materials.

MGI in the Automotive Sector

The automotive industry has been and continues to be a leader in the development and implementation of Integrated Computational Materials Engineering (ICME) tools, resulting in significant development cost savings and boosting competiveness for firms that have mastered these tools.[37] In an early example, Ford Motor Company researchers developed a suite of ICME software tools that captured extensive knowledge of aluminum casting technology, aluminum metallurgy, and mechanical behavior and product durability, enabling more rapid development of new products and casting processes. Following this ICME adoption, Ford Motor Company reported over a seven-to-one return on investment. (A cast aluminum Ford Duratech V6 engine block designed using Ford virtual aluminum castings ICME tools is shown below.)

The Materials Genome Initiative (MGI) provides a means to enhance and accelerate such developments. The continually increasing need to reduce the environmental impact of automobiles requires significant reductions in vehicle weights and major advances in powertrain technology. With the primary objective of accelerating new materials development, MGI will play an important role in ensuring that these needs are met.

To date, the automotive industry has mainly applied ICME tools for rapid, lower-cost product development using existing metal alloys, but similar tools can also be applied to new alloy designs. One of the first new alloy development programs resulting from MGI likely will be the rapid development of new cast aluminum alloys for automotive powertrain components. U.S. automotive companies, including General Motors and Chrysler, in collaboration with their suppliers and researchers at universities and national laboratories, have launched programs to develop cost-effective, cast aluminum alloys with significant improvements in elevated temperature properties, such as strength and resistance to cyclic fatigue loading. These alloys are expected to lead to reduced vehicle emissions by enabling higher exhaust gas temperatures and significantly reducing engine weight. New alloy demonstrations in running engines are expected within the next four to five years, a significant acceleration of the typical 20-year timescale for new materials and a mark of success for the techniques and approach to materials research and engineering MGI advocates.

---

[37] *Ibid.* 1.

Science and Technology
Grand Challenges

Technological advances for national security, human health and welfare, clean energy, infrastructure, and consumer goods are critical in ensuring a thriving Nation for future generations. The success of the Materials Genome Initiative (MGI) in providing a technological and research framework to accelerate the deployment of materials solutions in these sectors will require addressing a variety of cross-cutting challenges across both materials classes and application domains. Through two Grand Challenge Summits, organized in 2013 by the interagency Subcommittee for MGI, the scientific and engineering community explored several key materials classes and applications in which to apply the MGI approach. The summits held focused discussions on biomaterials, catalysts, correlated materials, electronic and photonic materials, energy storage materials, lightweight and structural materials, organic electronic materials, polymers, and polymer composites. [38] Summit participants included representation from academia, National and Federal laboratories, industry, and Federal agencies. These summits provided a communication venue across multiple groups to ensure that research, manufacturing, and commercial industry perspectives were considered as input for this strategic plan.

Summit participants were asked to identify grand challenges that would inspire and enable future MGI-related research to accelerate innovation and technology development across the materials and applications spectrum. Within each materials class or application domain, participants identified grand challenges that are, at present, still aspirational. As research progresses, a subset of these grand challenges is expected to become better defined and yield focus areas with quantifiable milestones for the MGI community.

Many of these grand challenges directly support national objectives in clean energy, national security, human welfare, infrastructure, and consumer goods. The selected materials classes and applications are shown in Table 1 and include indications of primary and secondary priorities within identified areas of national need.

---

[38] The materials classes selected for these summits are not intended to be comprehensive, nor to indicate that other materials classes are not MGI priorities. Future workshops to identify additional grand challenges may include, for example, ceramics, alloys for extreme environments, cements, energetic materials, and gas separation media.

Table 1. Materials Classes and Application Domains Included in Grand Challenge Summits and Their Relationship to National Needs

	National Security	Human Health and Welfare	Clean Energy Systems	Infrastructure and Consumer Goods
Biomaterials	○	●	○	●
Catalysts	○	●	●	●
Polymer Composites	●	●	○	●
Correlated Materials	●	○	●	●
Electronic and Photonic Materials	●	○	●	●
Energy Storage Systems	●	●	●	●
Lightweight and Structural Materials	●	●	●	●
Organic Electronic Materials	○	●	○	●
Polymers	○	●	○	●

● Primary   ○ Secondary

The summits generated a brief overview of the role and importance of each specified materials class or application as well as a corresponding list of the scientific or technical challenges facing the community that MGI could help solve. Several common or cross-sector themes emerged from the summits, including (1) support for the culture change needed to embrace the deeper integration of experiment and modeling at all stages of the materials development continuum, (2) integration of tools at multiple length and timescales, (3) access to and curation of data and material samples, (4) linking discovery and development with manufacturing processes, and (5) education in both simulation and experiment for the next generation workforce.

The remainder of this chapter comprises the output generated by the summits and approved by the session chairs to be representative of the opinions of the workshop participants.

Biomaterials

The field of biomaterials has undergone major transformations over the past two decades. Fifty years ago, the only materials used in biomedical applications were largely already known from other technology applications, including, for example, metals and polymers used to reconstruct diseased joints or replace segments of large blood vessels. Today, the field encompasses not only areas in which the primary objective is to repair human tissues, but also biomimicry, in which synthetic structures are created by imitating biological materials, and biological systems to synthesize useful materials. Biomaterials remain a multibillion-dollar industry that saves lives and enhances human welfare.

In the MGI context, four distinct directions should be pursued to benefit both national and global interests in health, energy, and sustainability: (1) bioactive biomaterials for regenerating human tissues and organs; (2) bioinspired materials that transduct energy the same way muscles do, self-assemble into hierarchical structures with currently unknown properties, repair themselves, or adapt to their environment; (3) biofabricated materials that involve harnessing biology to make materials, especially with new capabilities emerging for genetic manipulation of cells; and (4) materials to interface with biology for the discovery of new materials that can interrogate or modulate the functions of biological systems, such as bacteria or stem cells, in applications that include sensing, regeneration, drug discovery, or fuel production. These four areas are a rich source of new sustainable technologies for economic competitiveness. The following is a list of some of the MGI-relevant grand challenges for biomaterials:

- Develop theoretical and modeling tools across length and timescales.

- Accelerate the development of dynamic self-assembly of materials and harness biology for materials synthesis and fabrication.

- Design materials that form three-dimensional (3D) self-assembling functional objects with chemistry that mimics the fidelity of Watson-Crick pairing (i.e., a non-DNA DNA).

- Utilize bioactive materials for regenerative medicine.

- Create materials that control the functions of living systems (or vice versa).

- Develop strategies to obtain chemically sequenced synthetic polymers.

- Develop strategies to create emergent properties in materials.

- Develop tools for nondestructive structural characterization of biomaterials at varying scales to discover links to function.

Catalysts

A catalyst is a reactive material in which the active site as well as its working environment is critical to performance and selectivity of desired products. Catalysts are an enabling technology critical to many U.S. industrial sectors including energy, chemicals, and pharmaceuticals. For example, the development of a catalyst that splits water efficiently and cheaply on commercial scales would revolutionize the energy industry and significantly reduce carbon dioxide emissions. What follows are grand challenges that would enable the vision of significantly decreasing the time and cost involved in the discovery and deployment of new catalysts:

- Develop modeling tools that go beyond what fundamental theory (e.g., density functional theory) can do, reach longer length and timescales with higher accuracy, and represent complex environments and reaction networks.

- Enable better catalysis science by experimental and computational definitions of active sites and their functions, while accelerating applications.

- Develop advanced or new *in situ* spectroscopic and microscopic techniques for evaluating catalyst structure and properties under real operating conditions.

- Create and implement an open-access database for catalysts, catalytic rate, and thermochemical data.

- Create new synthesis strategies that enable catalyst designs, incorporate multiple functions defined at the molecular level, and can be applied at all levels from the laboratory through scale-up and commercialization.

- Develop tools to utilize thermodynamic and phase diagram information or data mining of literature to suggest appropriate synthesis techniques, conditions, and precursor materials.

- Establish materials and testing standards for evaluating and reporting catalytic performance (e.g., time of flight) and characterization protocols (e.g., surface area measurements), and verifying identification of materials.

Polymer Composites

Due to their highly specific mechanical properties, polymer composite materials originally were developed for aerospace applications. These materials now are experiencing rapid commercialization in other industries, including the automotive and sporting good sectors. Being able to tailor properties for specific applications through constituent selection and placement provides highly optimized components for product design. This ability to "design in" specific properties creates an exciting new opportunity to add multifunctionality to polymer composite materials, thus enabling unique product designs that efficiently combine mechanical, electrical, thermal, optical, and/or magnetic performance. What follows are the major scientific and technical challenges relevant to MGI identified in the polymer composite field:

- Image a 3,500 cubic centimeter (cm^3) cube of a composite component fully in 3D with resolution at the level of, for example, constituents, orientation, and distribution.

- Develop measurements and models to determine non-equilibrium, polymer molecular mass, and chemical functionality changes during cure in a 3D component.

- Develop an open, curated database of composite test and simulation data.

- Perform "reactive molecular dynamic simulations" in which chemical bonds are allowed to break and form as needed to predict ultimate properties.

- Quantitatively and more realistically describe microstructure by including variations in local stoichiometry, defect morphology and distributions, and composition gradients.

- Predict onset and propagation of damage with a specified confidence interval through accurate modeling.

- Capture all processing-relevant phenomena (including uncertainty) in multi-physics/chemistry kinetic models.

- Measure properties and their variations at all relevant length and timescales, from individual atoms to macroscale, using rapid experimental techniques.

- Model the evolution of residual thermal strain, particularly for the case of very high modulus carbon fibers.

Correlated Materials

Many recently discovered materials for new and emerging technologies have extraordinary properties that result from the interactions of electrons, which are part of the materials' atomic structure. Examples of these correlated electron materials include high-temperature superconductors, spintronic materials, magnetic materials, giant magnetoresistance materials, and topological insulators. Understanding and predicting the behavior of these materials require theory and models that go beyond simple consideration of electrons as non-interacting, single entities. MGI offers the potential for bringing these materials to the same level of predictability as conventional semiconductors, opening new opportunities for use of these materials in solutions to some of the Nation's major technological challenges. Specific grand challenges on the path to these goals include:

- Rapidly survey these materials using tools that incorporate correlation effects to produce trends in formation energies, structure, and excitations.

- Use multivariable optimization techniques to enable guided synthesis of new materials classes.

- Model correlated materials' structure and growth.

- Develop sub-10 nanometer (nm) device fabrication capabilities, looking toward a nano-3D printer in the long term.

- Model complex devices using system models that integrate from the nanoscale upward, bridging scales and methodologies.

- Integrate simulation and experiments, particularly at large user facilities where some experiments generate large, four-dimensional data sets.

- Create new devices by controlling correlated phenomena, taking advantage of opportunities in interface engineering in oxides, nanoscale control of electrochemistry, and defect engineering for nonlinear memory devices.

Electronic and Photonic Materials

Devices and components produced by the electronics and photonics industries are crucial to almost any application from national security to energy to human welfare. While the sophistication and scale of the electronics and photonics industries are exceptional, improvements to electronic and photonic materials, as well as to the manufacturing processes used to produce devices, are necessary to support continued performance improvements and domestic technology leadership. Successfully addressing the following grand challenges would support more rapid advancement in electronics and photonics and drive resulting improvements across a wide range of systems and applications:

- Predict excited states, transport, and non-equilibrium structures in electronic materials.

- Demonstrate highly accurate theories and methods for modeling electrical or optical properties of materials in structures smaller than 10 nm.

- Establish prediction models of full-device, emergent, or system properties using inputs from material properties, modes of integration, processing history, structural or defect attributes, and spatial or geometric features.

- Develop models and validate data to enable transition from bench-type design to design of a fabricated component with existing equipment.

- Implement tools that progressively validate, and render transparent, materials-centric databases (i.e., facilitating understanding rather than providing data).

- Model and predict the properties of a device, circuit, or electronic system at production scale using information only obtained at research scale.

- Model and predict the part-to-part variability of production devices as a function of material features and processing.

Energy Storage Systems

The need for reliable energy storage transcends boundaries separating private, governmental, and military sectors and is vital to the national well-being. Applications are numerous and broad; energy storage devices encompass massive and sessile equipment for factory and residential needs, as well as small, light, and portable devices for electric vehicles, medical devices, and other applications. Rapid and efficient charging and charge stability within the storage media are defining characteristics of advanced systems. The rate at which charge is released is an equally important characteristic with fast-release capacitors existing at one end of the spectrum, batteries at the other end, and supercapacitors in between. Understanding and manipulating the role of materials and interfaces in charge acceptance, transport, and release are driving research for all systems.

During the MGI Grand Challenges Summit, participants identified battery research as the most pressing and proposed the following grand challenges:

- Enable stable new battery systems with high-energy density by elucidating bulk and interfacial reaction mechanisms for all plausible electrolytes including solids. Establish this knowledge base for five volt systems within five years.

- Identify and quantify low-rate degradation mechanisms that determine long-term failure modes to speed the confident implementation of new materials and new battery system designs.

- Accelerate synthesis of new materials and their incorporation into battery systems by advancing the breadth and capability of prediction tools; specifically, emphasize computational tools for inorganic chemistry and informatics, as well as the ability to calculate Pourbaix-like diagrams that include kinetics.

Other specific goals also were proposed:

- Enable discovery and design of new metal anodes.

- Link inherent physical and electrochemical materials properties.

- Develop prediction and design tools that account for additives and trace impurities.

- Enable discovery and design of a nonflammable, yet high performance electrolyte.

- Enable more stable aqueous systems for three volt aqueous batteries.

Lightweight and Structural Materials

The automotive, aerospace, heavy machinery, shipbuilding, rail, home appliance, and construction industrial sectors contribute nearly a half-trillion dollars to the annual U.S. gross domestic product.[39] All of these sectors depend on improved and affordable lightweight and structural materials for product differentiation and economic competitiveness. The following are representative, aspirational goals that, if achieved, will provide significant advances in the ability to predictively model the continuum in lightweight and structural materials:

- Quantitatively predict the corrosion behavior of any metal alloy and predict its influence on properties.

- Demonstrate the ability to fully characterize the microstructure in one cm^3 of a complex engineering alloy within one week.

- Establish an integrated experimental and modeling approach to nondestructively map in 3D the full tensorial residual stress field in a component with 10 millimeter resolution over a volume of 10 cm^3, including depths up to one centimeter (cm), within one day.

- Develop a means for defining representative volumes for higher length scale experiments, modeling, and designs.

- Create, develop, and operate federated databases and database tools providing easy data access. Priority areas include: thermodynamics, kinetics, elastic constants, thermal expansion coefficients, crystal structure, electric and thermal conductivity, and plastic properties.

- Develop analytical tools for efficient extraction of process-structure-property linkages from large datasets that can be executed with desktop-scale computational resources.

Organic Electronic Materials

Numerous sources project that the carbon-based, printable, and flexible electronics industry could achieve an economic impact of $10 billion or more in the next several years, impacting industries such as lighting, displays, sensing, energy conversion and storage, medical diagnostics, biocompatible electronics,

---

[39] See data from the Bureau of Economic Analysis (available online at www.bea.gov).

environmental monitoring, and many others.[40] These materials enable not only new form factors (such as lightweight, flexible, or stretchable components), but also critical new processing methods such as direct printing. These capabilities allow short-run, customized electronic systems manufacturing with significantly reduced entry barriers when compared with conventional semiconductor fabrication. To benefit from this exciting technological opportunity, reliable, standardized, and easily manufactured components based on soft materials are required. Additionally, a much more detailed understanding of the process steps used to fabricate devices and their influence on thin-film material structure and device performance is an essential prerequisite for accelerating the development of this nascent industry and further broadening its scope. This needed understanding will follow from solutions to the following grand challenges:

- Predict molecular crystal structures and polymorphs.

- Characterize and model material properties and behavior at different magnitudes and combinations of length, time, and dimensionality scales, including grain structures and mesoscale crystal and amorphous domain distributions.

- Project device property evolution at the molecular scale.

- Create a liquid-phase manufacturing paradigm.

- Develop a comprehensive model for organic electronic-biological interfaces.

- Discover markers for performance instability.

Polymers

Polymers are ubiquitous, both in high-tech applications and everyday life; nearly all industrial sectors, including energy, transportation, aerospace, electronics, biotechnology, pharmaceutical, packaging, and water management, rely on polymeric materials for critical components or processing steps. All of these industries and many others would benefit significantly from the design, prediction, and development of advanced functional polymeric materials. While the polymer industry is currently dominated by oil-derived polyolefins, new polymeric molecules could, in principle, be created with intricate structures and multiple, simultaneous functionalities that approach and even surpass those encountered in biological systems. With the expansive parameter space for discovery, development of new materials must rely on an MGI-based strategy built on model prediction, targeted synthesis, and fast 3D time-dependent data analysis and interpretation. Summit participants proposed the following key challenges:

- Develop mesoscale models to predict equilibrium and non-equilibrium polymer structure, morphology, and properties (including rheology), and to design polymer processing strategies that couple structure and properties.

- Design the hierarchical structure of polymeric materials for functionality.

---

[40] For example: Das, R., and Harrop, P., 2013. *Printed, Organic & Flexible Electronics: Forecasts, Players & Opportunities 2013–2023*, IDTechEx. (available online at www.idtechex.com/research/reports/printed-organic-and-flexible-electronics-forecasts-players-and-opportunities-2013-2023-000350.asp)s.

- Develop strategies to characterize and interpret 3D structure and dynamics in real time.

- Develop strategies to identify, model, predict, and control the evolution of polymeric material properties over long timescales.

- Develop computer-enabled approaches to design responsive polymers for extreme environments.

Concluding Remarks

The Subcommittee on the Materials Genome Initiative (SMGI) developed this strategic plan to present the path forward for the Materials Genome Initiative. Drawing from the combined input of the Federal agencies involved in MGI and the broader academic and industrial materials science and engineering communities, the SMGI has defined the specific goals and near-term milestones that will lead to meeting the President's challenge to decrease the time and cost of bringing materials to market. The multifaceted approach described in this plan (enabling a paradigm shift in culture; integrating experiments, computation, and theory; facilitating access to materials data; and equipping the next-generation workforce) is essential to achieving success.

This plan's aim is to enable the MGI community, including both Federal and private stakeholders, to use these goals and milestones to drive and focus research and development efforts in the coming years. For example, the grand challenges presented, while not intended to be comprehensive, include many examples of scientific and technical roadblocks that MGI can address. Building a Materials Innovation Infrastructure and using it to address these and other technical hurdles will enable the materials science and engineering community to play a key role in developing solutions for some of the Nation's most pressing challenges in health and human welfare, national security, clean energy, and economic prosperity, including infrastructure and competitiveness in consumer products. MGI is both a strategic and cost-effective investment in materials research and development with potentially significant economic, technological, and scientific benefits.

Appendix A: Agency Interests and Emphasis Areas

In February 2012, the Subcommittee on the Materials Genome Initiative (SMGI) was constituted as part of the National Science and Technology Council (NSTC) Committee on Technology (CoT) to facilitate a coordinated effort across Federal agencies to identify policies for supporting the goals and implementing the recommendations outlined in the Materials Genome Initiative for Global Competitiveness (MGI) white paper. SMGI member agencies continue to fund materials science and engineering research and development (R&D) efforts in support of their agency missions and responsibilities while contributing expertise and advice in the capacity of the NSTC to further the broader national effort in accelerating discovery, development, and deployment of advanced materials. In this section, the agencies describe their individual interests in materials science R&D and MGI priorities.

Department of Defense

Department of Defense (DOD) leadership considers the increasing emphasis on Integrated Computational Materials Engineering (ICME) being promoted by MGI vitally important to affordability and long-term technological innovation for future warfighting systems. As a mission agency, DOD is uniquely positioned to target relevant engineering problems with multidisciplinary R&D efforts integrated along the full materials continuum from discovery through development, deployment, sustainment, and retirement of assets. At the foundational level, DOD invests in basic research to explore materials through first-principles calculations, development and quantification of processing-structure-property relationships, new experimental and characterization tools, and computational tools to include multiscale modeling capabilities. Maturation of this knowledge and the development of industry-ready tools are accomplished through applied research and advanced development funding, as well as support from the Small Business Innovative Research (SBIR) and Small Business Technology Transition Research (STTR) programs where appropriate. Working with materials suppliers and original equipment manufacturers to help guide research, DOD will leverage the important investments being made in manufacturing science and technology through the Manufacturing Technology (ManTech) programs to establish transition partnerships. This coordination will accelerate the confident implementation of advanced materials and systems. Leading by example, DOD researchers and performers will engage with students and colleagues to develop the culture and influence the training of the next-generation workforce to fully meet the goals of MGI.

DOD coordinates efforts through its Community of Interest for Advanced Materials and Processes and with the NSTC subcommittee established to build and coordinate this initiative. The Military Departments and DOD agencies (Components) are focusing investments on both meeting mission goals and making viable the promise of integrated computational materials design and processing. Reducing the time required for materials design and manufacturing has the potential to accelerate both use and value in critical DOD applications. DOD invests in (1) developing the fundamental tools needed for further accelerating advances in national materials capabilities; (2) establishing the communications infrastructure required to support the storing and sharing of the vast amount of theoretical, computational, and experimental data necessary to speed the discovery to deployment continuum; and (3) educating the next generation of scientists and engineers in the optimum use of these advanced tool sets and databases.

Examples of DOD programs and projects that support MGI include collaborative and complementary ICME-related efforts across the Components' research enterprises such as: (1) advancing the fundamental science of computational and experimental methods; (2) capturing understanding of processing-structure-property-performance relationships in tools linking materials scientists and engineers to component and system designers to accelerate confident materials implementation from discovery through sustainment; (3) identifying mathematical approaches within stochastic and statistical frameworks for multiscale materials modeling; (4) developing reduced-order descriptions of structure and models of microstructural evolution with better management of inhomogeneity and uncertainty; (5) generating and curating data sets, from quantum chemical topology through experimentally derived properties; (6) developing sophisticated electronic materials through multidisciplinary and multiscale modeling; (7) designing and developing new materials with predictable performance for extreme dynamic environments; (8) integrating validated physical models of reaction kinetics and transport into computational fluid dynamics codes as tools for the design of advanced electrochemical power-generation and storage devices; (9) integrating experiments and modeling to create deeper understanding and tools for the design and manufacturing of high energy–density capacitors and titanium powder-processed components; (10) incorporating residual stress considerations in the design and production of nickel-based superalloy turbine engine structures; (11) developing a digital design system for high-temperature polymer matrix composites; (12) developing advanced manufacturing capabilities through the Lightweight and Modern Metals Manufacturing Innovation Institute; and (13) accelerating certification of existing materials in new applications.

Department of Energy

The Department of Energy (DOE) has a leading role in MGI to advance research and software for the design of matter for energy-related applications such as energy storage and solar fuels; for topics of broader national impact that strongly overlap the portfolio for lightweight and high-temperature structural materials; and for functional materials, such as catalysts and photovoltaic, magnetic, and superconducting materials. Current DOE MGI activities are concentrated within the Office of Science under its Office of Basic Energy Sciences (BES), Office of Energy Efficiency and Renewable Energy (EERE), and Office of Fossil Energy (FE). In addition, there is a longstanding history of materials research for national security in DOE's National Nuclear Security Administration (NNSA) and significant applied materials research conducted in the focused technology programs of the Advanced Research Projects Agency- Energy (ARPA-E).

BES supports fundamental research in materials sciences and engineering, chemistry, geosciences, and physical biosciences to understand, predict, and ultimately control matter and energy at the electronic, atomic, and molecular levels, including research to provide the foundations for new technologies relevant to DOE's missions in energy, environment, and national security. BES's MGI activity, Predictive Theory and Modeling, focuses on research that will lead to new theory and modeling design paradigms, validated through experiments, which will enhance the rate of discovery of new or vastly improved materials, material systems, and chemical processes. Activities include the development of new software tools and data standards that will catalyze a fully integrated approach from material discovery to applications. Also included is research to advance *ab initio* methods for materials and chemical processes, providing user-friendly software that captures the essential physics and chemistry of relevant systems. Equally important is harnessing the power of modern experimental techniques, including (1) materials characterization at BES-supported user facilities for x-ray and neutron scattering, (2) advanced materials synthesis that builds on techniques at BES-supported nanoscale science user facilities and core synthesis science program, and (3) analysis of chemical processes including energy-relevant processes such as combustion and catalysis. The program supports software centers as well as single-investigator and small-group research activities.

EERE supports high-impact applied research and technology development for a broad range of energy efficiency and renewable energy applications, where high-performance materials and processes play an important role. MGI activities within EERE support materials R&D through the application of demonstrated computational and experimental tools, while emphasizing competitive and efficient manufacturing processes and considering the impacts of these processes and materials on meeting the engineering challenges of real-world systems. Examples include applying computational tools to deliver higher-performing carbon fiber composites from lower-cost feedstocks and lower energy–intensity processers, accelerating development of substitutional materials for rare earth elements (REE) in magnets and advanced alloys, and researching new lightweight, high-strength alloys and composites for energy-efficient structural systems. All of these efforts focus on enabling a wide range of cross-cutting technologies for use in industry, supporting energy-efficient and clean-energy products and applications. EERE-supported MGI efforts link competitive, scalable, and energy-efficient manufacturing and process R&D to controlling and improving material properties, such as through the use of ICME techniques and other investments.

FE supports, through our nation's laboratories and universities, the continued advancement of science and engineering focused on providing transformational fossil energy technology options to fuel the Nation's economy, strengthen security, and improve the environment. The MGI culture and approach is critical in accelerating the maturation of technologies that will allow the United States to use our fossil fuel resources efficiently, while minimizing environmental impacts and maintaining a global energy leadership role. Specifically, the FE portfolio is leveraging integrated, multiscale computational and experimental approaches in numerous activities, including the development of engineered materials for carbon capture, metal alloys for extreme environments, catalysts for gas conversion, and engineered-natural material systems relevant to carbon sequestration.

NNSA's responsibility to maintain U.S. nuclear deterrent capabilities requires both fundamental and applied science. Indeed, NNSA requirements for understanding both advanced computational methods and material performance under extreme conditions without nuclear testing frequently have led to developments in integrated computational materials science. In particular, NNSA's emphasis is on understanding the aging of materials ranging from polymers to actinides and understanding materials under extreme conditions, as well as all the fundamental work required to support these missions.

In support of a clean, secure, and affordable U.S. energy future, ARPA-E catalyzes and accelerates the transformation of scientific discovery into high-impact energy technologies that are too early in development for private-sector investment. Applied materials research plays a key role in many ARPA-E projects; ARPA-E performers in academia, small and large industries, and National and Federal laboratories will use the computational tools developed under MGI for advanced materials design and materials data analytics.

National Aeronautics and Space Administration

The National Aeronautics and Space Administration (NASA) provides MGI with the unique platform of continued understanding of materials for use on launch vehicles and other infrastructure that will be exposed to extreme environments. The goals, objectives, and priorities of MGI align with NASA's Technology Roadmap Areas 10: *Nanotechnology* and 12: *Materials, Structures, Mechanical Systems, and Manufacturing (MSMM)*, specifically in the area of computational material design. Determining the effects of mission-specific extreme environments on material performance and the revolutionary computational molecular and atomistic-based models required for the development of new composites, metallic alloys, and hybrid materials with unprecedented properties represents a long-term, but very high-payoff investment for NASA. This commitment will enable the Agency and the Nation to develop future-generation materials and build the essential physics-based understanding needed to ensure extreme reliability in complex systems.

NASA's Space Technology Mission Directorate (STMD) develops pioneering and cross-cutting technologies that enable multiple missions for internal and external stakeholders. By investing in high-payoff, transformational, and disruptive technologies that industry cannot tackle today, STMD matures the technology required for NASA's future missions in science and exploration and a vibrant space industrial base. Within the STMD portfolio, MGI is poised to play a vital role in materials, structures, and advanced manufacturing projects.

A major priority is to develop technologies that can reduce the time lag—currently about 20 years—between discovery and acceptance of a new material by the aerospace community. In addition, about $400 million is spent in the process of moving a material through the certification and acceptance process. The revolutionary materials needed to achieve the goals described above have yet to be developed using existing (i.e., heuristic and trial-and-error) methodologies; new approaches are needed for the design, development, manufacture, certification, and sustainment of lightweight materials and structures. NASA's long-range MGI vision is to include materials and manufacturing as full-fledged elements in the digital design process. The objective of the MGI project is to deliver computationally guided materials design for thermal protection systems (TPS), structural materials, and smart materials, as well as relational databases for superalloys, ceramic matrix composites (CMCs), and multifunctional materials. The project goals will be to (1) enable cross-center modeling efforts for emerging material systems, including multifunctional materials for aerospace applications; (2) define the path for compressed materials maturation and insertion through multiscale modeling to reduce materials testing and shorten the iterative cycle for materials optimization; and (3) give materials designers the capability to assess trade-offs between selected material properties of interest and rapid prototyping. Additionally, NASA will coordinate with other efforts by SMGI member agencies to spur U.S. manufacturing by reducing the time to market for emerging material systems. NASA will align its activities with materials development areas of interest in NASA's Technology Areas 10 and 12 Roadmaps, *Nanotechnology* and *MSMM*, respectively.

National Institute of Standards and Technology

The missions of MGI and the National Institute of Standards and Technology (NIST) are tightly aligned. NIST promotes U.S. innovation and industrial competitiveness by advancing measurement science, standards, and technology in ways that enhance national economic security and improve quality of life. MGI addresses precisely these mission elements by providing the means to reduce the cost and development time of materials discovery, optimization, and deployment. Both missions are driven by industrial competitiveness, with the creation of a Materials Innovation Infrastructure as the means to this end.

Given NIST expertise in the integration, curation, and provisioning of critically evaluated data, NIST has assumed a leadership role within MGI. To foster widespread adoption of the MGI paradigm both across and within materials development ecosystems, NIST is establishing essential data exchange protocols and the means to ensure the quality of materials data and models. These efforts will yield the new methods, metrologies, and capabilities necessary for accelerated materials development. NIST is working with stakeholders in industry, academia, and government to develop the standards, tools, and techniques enabling acquisition, representation, and discovery of materials data; interoperability of computer simulations of materials phenomena across multiple length and timescales; and quality assessments of materials data, models, and simulations.

Internally, NIST is conducting several path-finder projects to develop key aspects of the Materials Innovation Infrastructure, expose challenges in the infrastructure's construction, and serve as exemplars for the broader MGI effort. These efforts include pilot projects to develop superalloys and advanced composites, both of which are new, energy-efficient materials for transportation applications. NIST's Material Measurement Laboratory coordinates these activities in partnership with the NIST Information Technology Laboratory, with broad participation across the Institute. To support this effort, NIST is pioneering curated repositories of materials data and models that result from research funded by a DOE EERE program in lightweight automotive materials. NIST expects to extend this approach to other agencies, both through direct partnerships and the dissemination of best practices.

In order to achieve these ambitious goals, NIST has dedicated $5 million per year for up to 10 years to fund a Center of Excellence in Advanced Materials. In December 2013 the co-operative agreement between NIST and a Chicago-based team, the Center for Hierarchical Materials Design (CHiMaD), was announced. The new center will focus on developing the next generation of computational tools, databases and experimental techniques to enable "Materials by Design," one of the primary goals of the Administration's Materials Genome Initiative (MGI). CHiMaD will focus these techniques on a particularly difficult challenge, the discovery of novel "hierarchical materials." Hierarchical materials exploit distinct structural details at various scales from the atomic to the macroscale to achieve special, enhanced properties.

For fiscal year 2015, the Administration has proposed broadening the NIST effort, with investments in critical MGI infrastructure. Priority areas include deepening NIST's investment in measurement science and data infrastructure for advanced materials, pursuing the development of co-designed advanced computational and experimental techniques, and analytical methods to capitalize on the emerging discipline of "big data" for materials applications.

National Institutes of Health

The National Institutes of Health is the primary Federal agency for conducting and supporting medical research. The NIH mission is to seek fundamental knowledge about the nature and behavior of living systems and the application of that knowledge to enhance health, lengthen life, and reduce the burdens of illness and disability. Toward these ends, NIH leadership realizes that advances in materials and in particular biomaterials have the potential to make valuable contributions to biology and medicine, which in turn could contribute to a new era in healthcare. The Federal agencies' R&D investments, for example, have resulted in advanced materials, tools, and instrumentation that can be used to study and understand biological processes in health and disease. NIH-supported R&D efforts, in particular, are bringing about new paradigms in the detection, diagnosis, and treatment of common and rare diseases, resulting in new classes of therapeutics and diagnostic biomarkers, tests, and devices.

NIH supports the Materials Genome Initiative by stimulating R&D in biomaterials development through both intramural and extramural funding. For more information on specific topics funded by NIH, please visit the NIH Research Portfolio Online Reporting Tool at www.report.nih.gov. NIH institutes also support large center grants, program grants, and small businesses whose technologies or products are licensed or currently undergoing Phase I–III clinical trials.

National Science Foundation

The National Science Foundation (NSF) supports fundamental scientific and engineering research that leads to discoveries promoting national health, prosperity, and welfare. New and advanced materials are critical in facets of all these national needs; thus NSF is excited to participate in MGI through its program, Designing Materials to Revolutionize and Engineer our Future (DMREF). MGI recognizes the importance of materials science and engineering to the well-being and advancement of society and aims to "deploy advanced materials at least twice as fast as possible today, at a fraction of the cost." As a national initiative, MGI integrates all aspects of the materials continuum, including materials discovery, development, property optimization, systems design and optimization, certification, manufacturing, and deployment. Integration of materials theory, advanced computational methods and visual analytics, data-enabled scientific discovery, and innovative experimental techniques is critical for the necessary revolution in this approach to materials science and engineering. NSF will promote this integration through its DMREF program.

Consistent with its focus on fundamental research, NSF is interested in activities that accelerate materials discovery and development by enhancing the knowledge base and understanding needed to progress toward designing and making materials with a specific and desired function or property from first principles, an approach often called "matter by design." The complexity and challenges addressed by MGI require a transformative approach to discovering and developing new materials, optimizing and predicting material properties, and informing material system design. Accordingly, research supported by DMREF must be a collaborative and iterative process whereby computation guides experiments and theory, while experiments and theory inform computation. Through the promotion of this collaborative and iterative process, NSF activities will enable realization of this strategic plan's first goal: to facilitate a paradigm shift in materials science and engineering research, development, and deployment methods. To further support the achievement of this goal, NSF encourages new approaches to materials education that provide students with the knowledge and experiences needed to actively participate in this new approach to materials discovery.

Research funded through DMREF seeks to advance fundamental understanding of materials across length and timescales, thereby elucidating the effects of microstructure, surfaces, and coatings on the properties and performance of engineering materials. The ultimate goal is to control material properties through design via the establishment of interrelationships among composition, processing, structure, properties, performance, and process control, all validated and verified through measurements and experimentation. Required new capabilities include: (1) methods for creating and characterizing materials; (2) theoretical constructs for understanding materials phenomena and properties; (3) data analytics tools and statistical algorithms; (4) advances in predictive modeling that leverage machine learning, data mining, and sparse approximation; (5) data infrastructure that is accessible, extensible, scalable, and sustainable; and (6) collaborative capabilities for managing large, complex, heterogeneous, distributed data supporting materials design, synthesis, and longitudinal studies.

NSF initiated DMREF with awards in fiscal year 2012 and continues to support the program through well-coordinated activities involving the Directorates of Mathematical and Physical Sciences (MPS), Engineering (ENG), and Computer and Information Science and Engineering (CISE). Within MPS, the Divisions of Chemistry (CHE), Materials Research (DMR), and Mathematical Sciences (DMS) participate in DMREF. The ENG Divisions of Civil, Mechanical, and Manufacturing Innovation (CMMI); Electrical, Communication and Cyber Systems (ECCS); and Chemical, Bioengineering, Environmental, and Transport Systems (CBET) also participate. All CISE divisions engage in the DMREF initiative.

U.S. Geological Survey, Department of the Interior

Although MGI focuses mostly on the middle of the materials lifecycle—development of materials for manufacturing—there are important considerations on both the front and back ends: (1) discovery and processing of raw materials; (2) supply risk and materials flow; (3) tracking and fingerprinting resources such as conflict elements/minerals (e.g., diamonds, Coltan [niobium- tantalum mineral], and gold); and (4) recycling and disposal of materials. The U.S. Geological Survey (USGS) has extensive research activities in all these fields but especially in the first two. For example, USGS is the main source of Federal information on discovery, assessment, and production of mineral resources, which includes how and where to find any element in the periodic table that might be used in MGI research. An essential realization is that resources start in the Earth and not in a laboratory or manufacturing plant. These things are interconnected. For example, when developing a new material or process, knowing the availability of the required resources is important. Metals like gold, platinum, and REE have many wonderful properties but also potential supply restrictions, both natural and political. Thus, consideration of the discovery part of the materials lifecycle could influence the research and fabrication pathway. Similarly, knowledge of new research directions, such as cobalt in certain nanotechnologies, could influence future USGS research directions on ore discovery and assessment.

Appendix B: Related Federal Activities

Manufacturing

The Materials Genome Initiative (MGI) was launched by the President at the same time as the Advanced Manufacturing Partnership (AMP), a partnership across government, industry, and academia to identify the most pressing challenges and transformative opportunities to improve technologies, processes, and products across multiple manufacturing industries. Related activities across MGI, AMP, and other manufacturing initiatives illustrate the strong link between MGI and the Administration's efforts to revitalize the American manufacturing sector. Work through MGI will provide cutting-edge computational software, databases, and associated instrumentation that will give domestic manufacturing a competitive advantage, reducing the time required to introduce new materials and products, and to safely introduce modified materials into existing products.

Consistent with the President's vision for a National Network for Manufacturing Innovation, the Administration announced open competitions in 2013 for three new Manufacturing Innovation Institutes to join the existing National Additive Manufacturing Innovation Institute. One of these new institutes will be managed by the Department of Energy (DOE) and dedicated to the development of wide bandgap semiconductor power electronic devices. Another will be run out of the Department of Defense's Office of Naval Research with a specific materials focus on "Lightweight and Modern Metals Manufacturing," a rich area of research within the MGI member agencies. More recently, the President announced a new competition to establish an Advanced Composites Manufacturing Innovation Institute, the first of four competitions for new manufacturing innovation institutes to be launched in 2014.

In 2013, DOE launched a Clean Energy Manufacturing Initiative designed to increase U.S. manufacturing competitiveness in the production of clean-energy products and to boost U.S. manufacturing competitiveness across the board by increasing energy productivity. This DOE initiative encompasses several activities that MGI can leverage to accelerate the manufacture of clean-energy-related materials, including funding opportunity announcements for manufacturing research and development (R&D), as well as the development of new partnerships bringing together many sectors, including public and private industry, universities, think tanks, and labor leaders.

Open Access

The materials science and engineering community, and by extension MGI, will be beneficiaries of the Administration's movement toward open access of federally funded research data. In a February 22, 2013 memo, Office of Science and Technology Policy (OSTP) Director Dr. John Holdren directed Federal agencies with more than $100 million in R&D expenditures, including those agencies involved in MGI, to develop plans for making the published results of federally funded research freely available to the public within one year of publication. The memo also requires researchers to better account for and manage the digital data resulting from federally funded scientific research. Related efforts to develop a data infrastructure that supports curation, storage, and access to materials science research data will build on the ongoing work of these agencies as they develop policies to meet the directions laid out in the memo.

Other Federal Initiatives

Over the last several decades, there has been significant Federal investment in new experimental processes and techniques for designing advanced materials. MGI works to leverage existing Federal investments through the use of computational capabilities, data management, and an integrated approach to materials discovery, development, and deployment.

MGI builds on the materials characterization and synthesis capabilities developed through the National Nanotechnology Initiative (NNI). The ability to control synthesis and characterize the chemistry and structure of materials at the nanoscale provides the foundation for experimental expertise that must be merged with theoretical, modeling, and computational tools to realize the vision of MGI. In addition, the Nanotechnology Knowledge Infrastructure Signature Initiative strives to stimulate the development of models, simulation tools, and databases that enable predictions of nanoscale material properties. This signature initiative directly links MGI and NNI activities and creates an opportunity to leverage the successes and lessons learned by each as they strive to implement predictive tools for materials production and manufacturing.

MGI also has been coordinating with the Networking and Information Technology Research and Development Program (NITRD), a multi-agency program to provide R&D foundations for continued U.S. leadership in advanced networking, computing systems, software, and associated information technologies. The NITRD Big Data Senior Steering group works to identify current Big Data R&D activities, such as MGI, across the Federal Government and offer coordination opportunities.

Through NITRD, MGI will be able to take advantage of Federal investments to improve the ability to extract knowledge and new information from large and complex data collections. There is no exact estimate of how much materials science and engineering data exists in individual laboratories and companies presently; once the materials data infrastructure envisioned by MGI begins to take shape, an increasing amount of both new and archival data may be made publicly available. This level of increased data-handling capacity will enable new research avenues not previously envisioned and accelerate the pace of discovery and innovation.

Interagency Coordination

The Subcommittee on the Materials Genome Initiative (SMGI) was established in 2012 under the National Science and Technology Council's (NSTC) Committee on Technology (CoT) to advise and assist NSTC and OSTP on policies, procedures, and plans related to Federal activities in support of the goals of MGI. SMGI is designed to facilitate a coordinated effort across Federal agencies to identify policies for supporting the goals and achieving the vision of cutting in half the time and cost of bringing new materials to market.

SMGI organizes workshops or other interagency activities that inform the Federal Government's decision making process on advanced materials. Each agency participating in MGI is represented on SMGI.

Executive Office of the President

Representatives from the Executive Office of the President (EOP) participate in MGI activities to ensure that MGI implementation is coordinated and consistent with government-wide priorities. OSTP is the primary point of contact.

OSTP is responsible for advising the EOP on matters related to science and technology and supports coordination of interagency science and technology activities. OSTP administers NSTC, and this arrangement provides EOP-level input on and support for various MGI activities.

Appendix C: Acronyms and Abbreviations

3D	three dimensional
AIM	Accelerated Insertion of Materials program
AMP	Advanced Manufacturing Partnership
ARPA-E (DOE)	Advanced Research Projects Agency–Energy
BES (DOE)	Office of Basic Energy Sciences
CBET (NSF ENG)	Chemical, Bioengineering, Environmental, and Transport Systems Division
CHE (NSF MPS)	Chemistry Division
CISE (NSF)	Computer and Information Science and Engineering Directorate
CMC	ceramic matrix composite
CMMI (NSF ENG)	Civil, Mechanical and Manufacturing Innovation Division
CoT	Committee on Technology
DARPA	Defense Advanced Research Projects Agency
DOD	Department of Defense
DOE	Department of Energy
DMR (NSF MPS)	Division of Materials Research
DMREF (NSF)	Designing Materials to Revolutionize and Engineer our Future program
DMS (NSF MPS)	Division of Mathematical Sciences
ECCS (NSF ENG)	Electrical, Communications and Cyber Systems Division
EERE (DOE)	Office of Energy Efficiency and Renewable Energy
ENG (NSF)	Engineering Directorate
EOP	Executive Office of the President
FE (DOE)	Office of Fossil Energy
FEP	foundational engineering problem
GCDP (NASA STMD)	Game Changing Development Program
ICME	Integrated Computational Materials Engineering
LGPS	$Li_{10}GeP_2S_{12}$
ManTech (DOD)	Manufacturing Technology programs
MGI	Materials Genome Initiative
MPS (NSF)	Mathematical and Physical Sciences Directorate
MSMM (NASA)	Materials, Structures, Mechanical Systems, and Manufacturing Roadmap Area 12
NASA	National Aeronautics and Space Administration
NCNR	NIST Center for Neutron Research

NIST	National Institute of Standards and Technology
NITRD	Networking and Information Technology Research and Development Program
NNI	National Nanotechnology Initiative
NNSA (DOE)	National Nuclear Security Administration
NSTC (OSTP)	National Science and Technology Council
NSF	National Science Foundation
OSTP	Office of Science and Technology Policy
PSED	Predictive Science and Engineering Design
PMC	polymer matrix composites
R&D	research and development
REE	rare earth elements
SBIR	Small Business Innovation Research
SOFC	solid-oxide fuel cell
SMGI (NSTC CoT)	Subcommittee on the Materials Genome Initiative
STEM	science, technology, engineering, and math
STMD (NASA)	Space Technology Mission Directorate
STTR	Small Business Technology Transfer Research
TPS	thermal protection system
USGS	U.S. Geological Survey

Appendix D: Summary of Public Comment

SMGI received 36 comments in response to a request for public comment on a draft version of this plan during the period from June 20 to July 21, 2014. The majority of the input came from the academic community, with additional submissions from industry, scientific societies, national laboratories, and other stakeholders. Common themes that emerged in the comments included: clarity on the breadth of disciplines and stakeholders that contribute to materials science research and development, from the discovery stage through manufacturing of a product; the importance of uncertainty quantification; sustainability and reuse of materials; the role of industry in MGI; and clarification on the scope and origin of the grand challenges identified for nine materials classes and application domains. SMGI is deeply appreciative of the thoughtful replies and suggestions provided by the community, which led to improvements in the final version of this document.

www.ingramcontent.com/pod-product-compliance
Lightning Source LLC
Chambersburg PA
CBHW050752180526

45159CB00003B/1438